Springer Theses

Recognizing Outstanding Ph.D. Research

For further volumes:
http://www.springer.com/series/8790

## Aims and Scope

The series "Springer Theses" brings together a selection of the very best Ph.D. theses from around the world and across the physical sciences. Nominated and endorsed by two recognized specialists, each published volume has been selected for its scientific excellence and the high impact of its contents for the pertinent field of research. For greater accessibility to non-specialists, the published versions include an extended introduction, as well as a foreword by the student's supervisor explaining the special relevance of the work for the field. As a whole, the series will provide a valuable resource both for newcomers to the research fields described, and for other scientists seeking detailed background information on special questions. Finally, it provides an accredited documentation of the valuable contributions made by today's younger generation of scientists.

## Theses are accepted into the series by invited nomination only and must fulfill all of the following criteria

- They must be written in good English.
- The topic should fall within the confines of Chemistry, Physics and related interdisciplinary fields such as Materials, Nanoscience, Chemical Engineering, Complex Systems and Biophysics.
- The work reported in the thesis must represent a significant scientific advance.
- If the thesis includes previously published material, permission to reproduce this must be gained from the respective copyright holder.
- They must have been examined and passed during the 12 months prior to nomination.
- Each thesis should include a foreword by the supervisor outlining the significance of its content.
- The theses should have a clearly defined structure including an introduction accessible to scientists not expert in that particular field.

David D. O'Regan

# Optimised Projections for the Ab Initio Simulation of Large and Strongly Correlated Systems

Doctoral Thesis accepted by
The University of Cambridge, UK

Springer

*Author*
Dr. David D. O'Regan
Cavendish Laboratory
TCM Group
University of Cambridge
JJ Thomson Avenue
Cambridge, CB3 0HE
UK
e-mail: ddo20@cam.ac.uk

*Supervisor*
Prof. Mike C. Payne
Cavendish Laboratory
TCM Group
University of Cambridge
JJ Thomson Avenue
Cambridge, CB3 0HE
UK
e-mail: tcmadmin@phy.cam.ac.uk

ISSN 2190-5053
ISBN 978-3-642-23237-4
DOI 10.1007/978-3-642-23238-1
Springer Heidelberg Dordrecht London New York

e-ISSN 2190-5061
e-ISBN 978-3-642-23238-1

Library of Congress Control Number: 2011936135



*Cover design:* eStudio Calamar, Berlin/Figueres

Printed on acid-free paper

Springer is part of Springer Science+Business Media (www.springer.com)

*To my beloved parents*

# Supervisor's Foreword

Density functional theory is remarkable. By searching over the single particle electron density alone, in principle, it provides the exact quantum mechanical ground state energy of a given system and the corresponding exact ground state electron density. It achieves this incredible feat via the exact density functional, which, for any input electron density, outputs the sum of the kinetic, Hartree and exchange-correlation energies for the ground state many-body wavefunction which generates this input electron density. As we add the final electrons to a semiconductor when filling the valence bands, the density functional tells us that the change in energy on adding each successive electron is almost constant. However, when we add an extra electron, which has to enter the conduction band, the exact density functional tells us that the change in energy must differ, discontinuously, from its previous value. The difference in these energies is, in fact, the band gap of the semiconductor and exact density functional theory reproduces its value precisely even though the single particle electronic density changes by an infinitesimally small amount each time an electron is added to the system. Additionally, although the contribution to the single particle electron density from a given set of orbitals might equate to four and a half electrons, for example, the exact density functional captures the knowledge that electrons cannot be divided and that this occupancy can only occur because the true many-body wavefunction is a superposition of configurations, perhaps one which puts four electrons in this set of orbitals while another represents five electrons. In situations like this, the exact density functional is capable of precisely determining the energies associated with this many-body wavefunction.

Unfortunately, we are not as clever as the true density functional. Our crude approximations of the density functional break some of the physical constraints on the true many-body wavefunction, with a consequent detrimental effect on the predicted energies and densities. For example, all of the available density functionals get band gaps wrong, often badly underestimating them, and numerous approximate functionals quite happily place non-integer numbers of electrons on particular sites. It is my view that no density functional that we may create will ever overcome all of these problems simultaneously, although it is often possible to overcome one shortcoming using an approximate functional designed for that purpose. However, while such a functional will, by construction, give better results for the targeted property, it may then give worse predictions of other properties than standard functionals and it is thus in no real way closer to the exact density functional.

Given the complexity of many-body wavefunctions it is, perhaps, remarkable that available density functionals work as well as they do—often predicting physical properties to within an accuracy of a few percent. Furthermore, density functional theory allows us to perform predictive calculations on systems containing many thousands of atoms, while we can only compute many-body wavefunctions for a handful of electrons. One approach to alleviate the shortcomings of the available density functionals is simply to insert the physics that is missing in these approximate functionals. For instance, it has now become very common to add an explicit van der Waals interaction between the atoms as an additional contribution to the total energy in a density functional theory calculation. Another widely used approach is the so called DFT + $U$ method whereby a Hubbard $U$ interaction is added to reproduce the physics of strongly correlated localised electronic orbitals. The weakness of previous implementations of DFT + $U$, in which the occupancy of the orbitals is constrained to be an integer number of electrons, was that the results depended on the choice of the projectors used to determine the occupancy of the localised orbitals. This thesis presents a method in which these projectors may be determined self-consistently during the DFT + $U$ calculation, thus providing an approach to overcome this weakness in previous implementations. This approach has been implemented in the linear scaling density functional code ONETEP and is shown to retain the linear scaling of computational cost with system size. This thesis contains applications of this technique to bulk nickel oxide, ligated iron porphyrins of biological interest and the copper phthalocyanine dimer, as well as scaling tests on nickel oxide nanoclusters containing over 7,000 atoms.

In order to develop the projector self consistent DFT + $U$ methodology, it was necessary to master the full mathematical complexities of tensorial calculus in the context of electronic structure calculations. This thesis contains a detailed exposition on the use of nonorthogonal orbitals, the construction of contracted tensorial invariants, energy minimisation algorithms on curved spaces and the Christoffel symbol corrections needed to ensure that the density matrix retains its idempotency, to first order, as the functions in which it is expanded are updated. This thesis provides a very detailed, yet readable, account of these issues and could become the standard reference on this topic for the electronic structure community.

Many technological materials rely on strongly correlated electronic systems for their functional properties and atoms that host strongly correlated electronic orbitals are found in the active sites of many proteins. DFT methods have usually struggled to describe such systems accurately and the results of DFT + $U$ studies have fundamentally depended on the set of projectors used in such calculations. As a result of the work presented in this thesis, we are moved a step closer to the accurate and routine description of such systems using first principles quantum mechanical approaches.

Cambridge, June 2011 Prof. Mike C. Payne

# Acknowledgments

This dissertation comes, ostensibly, as a culmination of one man's labours over the past three years. That it is but, as Donne wrote, "No man is an island, entire of itself" and it is a pleasure take some time here to thank those organisations and individuals who have contributed to this work and to my life over this period.

My research has been generously supported by the UK Engineering and Physical Sciences Research Council and the National University of Ireland. The Cambridge HPCS and, via the UK Car-Parrinello Consortium, the UK National Supercomputing Service HECToR have provided much of the required computing resources. Pembroke College has provided travel grants, much pastoral support and a welcoming home for a good part of my time in Cambridge. My sincere thanks extends to these organisations for their assistance.

The design of this dissertation is derived from a style package due to Thomas Fink and Robert Farr, though any inconsistencies in the layout are purely of my own making.

The Thomas Young Centre at Imperial College London has allowed me to make frequent visits to the Mostofi group, which has been my academic home away from Cambridge for the past three years. I would like to warmly thank my friends at Imperial for their excellent welcome and all that they have taught me.

One could not ask for a friendlier and more stimulating work environment than the TCM group at Cavendish Laboratory; I have very much landed on my feet in that sense. It was a great privilege to share an office with Jamie Blundell and John "Maestro" Biggins for three years; I think that the support shared there more than outweighed the ample distractions! All members of TCM have enriched my experience in some way, but I would like to particularly acknowledge Andrew Morris, Hatem Helal, Alex Silver, Robert Lee, Jonathan Edge, Gareth Griffiths, Priyanka Seth, Danny Cole, Patricia Silas, Mark Robinson, Sîan Joyce, Professor Mark Warner and Professor David Khmelnitskii for their support and advice. I am also grateful to the regulars at the TCM DFT meetings for their help and enthusiasm. Last, but by no means least, I extend my sincerest thanks to Tracey Ingham and Michael Rutter for unstinting generosity with their time and expertise.

Much of my efforts have centred around the ONETEP code and it is my pleasure to acknowledge all of the developers and fellow contributors to this great work for their patience and professionalism. In particular, I thank Simon Dubois, Peter Haynes and Chris-Kriton Skylaris, who also kindly proof-read Chaps. 5 and 6, for stimulating discussions and suggestions. Nicholas Hine deserves a very special mention and thanks for a great deal of time spent guiding me; I have learned a huge amount from him in many matters and it is doubtful whether linear-scaling could be achieved for DFT + $U$ without his help.

Professor Charles Falco, Professor Stephen Fahy and Dr. Michel Vandyck have been my academic mentors prior to postgraduate study, and without their invaluable encouragement I might not have commenced this work at all. Dr. Jonathan Yates and Professor Matteo Cococcioni have helped to direct my research via stimulating discussions. My examiners, Professor Nicola Marzari and Professor Emilio Artacho, offered some very helpful advice and comments on the manuscript. I am very grateful.

The friends I have made in Cambridge have got me through this process and made it possible, we've shared many ups and downs. I particularly mention Katia Shutova, Michelle Rigozzi, Kelsey Edwardsen, Matt Smith, Taylor Hathaway-Zepeda, Elizabeth Dearnley, Emma Firestone, Matthias Wivel, Robin Payne, Peter Evan and Krishnaa Mahbubani for taking care of me, I cannot thank them enough.

I fear that my exile might be made permanent if I neglected to thank Linda Mason, Jennifer Lavin, Niall Johansson, Aoife FitzGibbon O'Riordan, David O'Farrell, Sinéad Rose and David Sheehan for their loyal friendship. I promise I will try harder to stay in touch. Thank you too to all at Munster Vintage Motor Cycle and Car Club.

The proximity of my dear friend Shane Mansfield has been a very great comfort to me. Thánamair abhus anso le chéile sa bhád agus, le cúnamh Dé, is sa chaoi chéanna go bhfillfimíd thar n-ais aríst lá éigin.

My teacher, advisor, critic, counsellor and friend; the game would be lost completely if it were not for the unwavering guidance and generosity of Dr. Arash Mostofi. My obligation to Arash is great, I thank him a thousand times.

I thank my supervisor, Professor Mike Payne, for his excellent advice over these years, long hours spent proof-reading and straight answers when I needed them most. Mike has been an enthusiastic advocate at important moments and I'm very grateful indeed.

The love and kindness shown by Florence Paul over these years has truly kept me going. I hope to be repaying it for a very long time.

Finally, I would like to thank all of my family for their love and minding since day one. My new goddaughter, Isabel, has brightened up a very cold winter spent in writing. I would be lost without my wonderful sister, Aoife, and my beloved Mother and Father, Bernice and John, to whom all of this is dedicated.

# Contents

# Chapter 1
# An Introduction to Linear-Scaling Ab Initio Calculations

Atomistic modelling is a powerful tool that allows numerical experiments to be conducted which may be used to predict the properties of new materials, or known ones under novel user-defined conditions, and to test the validity of physical models against experiment.

Density functional theory (DFT) is, without a doubt, the most successful method currently available for making quantitative theoretical predictions of the properties of real materials. The popularity of atomistic simulations using DFT is owed to its capability of accurately and reliably reproducing the ground-state properties of many quantum-mechanical systems *ab initio*, that is provided only with a set of reasonably well-controlled approximations and the atomic positions. Those positions may, if so required, also be altered according to quantum-mechanical forces.

Physical attributes determined by the electronic ground-state and reproducible using DFT include: equilibrium geometric structures and any magnetic ordering; phase transitions between structures; cohesive binding energies; elastic moduli; transfer of charge and magnetisation between species; local electric and magnetic moments; polarisabilities and susceptibilites in the low and high-frequency limits; and potential energy isosurfaces [1]. Moreover, the results of a ground-state DFT calculation often serve as an excellent starting point for sophisticated methods for describing excited-state properties. These include the *GW* approximation [2–6], Dynamical Mean Field Theory [7–10] and time-dependent Density Functional Theory (TDDFT) [11, 12]. We refer the reader to excellent reviews on atomistic modelling using DFT in Refs. [1, 2, 13, 14].

## 1.1 The Challenges of Spatial and Electronic Complexity

A number of factors, unfortunately, currently limit the applicability of atomistic modelling with DFT and hence its usefulness for guiding or stimulating experimental innovation. Foremost among these is the issue of computational expense: the realistic study of many systems requires the explicit treatment of system sizes

D. D. O'Regan, *Optimised Projections for the Ab Initio Simulation of Large and Strongly Correlated Systems*, Springer Theses, DOI: 10.1007/978-3-642-23238-1_1,

(or simulation times, if ionic dynamics are of importance) which are beyond the reach of currently available computing technology. This is particularly true when conventional algorithms for solving for the ground-state density, which scale cubically with the number of atoms, are used. Irrespective of ever-increasing computational resource availability, it is only by using algorithms for which the effort increases linearly with system size that we may routinely bring first-principles simulation to bear on many complex systems which are pertinent to the technological, environmental and medical challenges of the twenty-first century.

Additionally, one finds that for a large class of systems of immediate technological or biological interest, particularly some of those comprising open-shell first-row transition metal or lanthanoid ions and categorised as "strongly-correlated", DFT in its most commonly practiced forms may perform very poorly, even qualitatively disagreeing with experimental observation in certain respects. In strongly-correlated systems, simple mean-field approximations for the interactions between electrons exceed their regime of applicability and more sophisticated techniques are needed to augment their description. A different, but not altogether unconnected issue is the intrinsic unsuitability of ground-state DFT for describing excited-state properties such as absorption spectra, multiplet splittings, non-equilibrium transport etc., when it is often these properties which are most easily observed experimentally.

In this dissertation, we present computational methodology to tackle the obstacles of large system size and strong correlation effects simultaneously. Examples of comparatively large systems where strong electron interaction effects play a role, and of the type that we have in mind, are shown in Fig. 1.1. From left to right, these are: a catalytic rare-earth oxide surface in a partially reduced multiple-valence state, which has a high density of sites of strong electron correlation; the crystal structure of a well-studied molecular magnet [15], whose antiferromagnetically coupled iron centres (green prisms) harbour strong correlations; and a B-DNA structure intercalated with two artificial base pairs mediated by $Cu^{2+}$ ions (green spheres) [16].

We work in the framework of a linear-scaling implementation of DFT in order to tackle the issue of system size, which allows us to demonstrate computational performance tests on a strongly-correlated oxide system of over 7,000 atoms. Much of our theoretical investigation focuses on the optimisation of nonorthogonal representations of single-particle states, which we show to be necessary for accurate, fully ab initio simulations of very large systems, and we derive a number of new analytical results in this area. For the study of strongly correlated systems, we employ the popular and efficacious DFT + $U$ method. We detail a linear-scaling implementation of DFT + $U$, rigorously generalise it to the case of nonorthogonal projections and offer a novel ab initio method for their optimisation.

## 1.2 Outline of Dissertation

We begin the dissertation with a brief introduction to the construction of linear-scaling ab initio methods and to the spatially localised representations which are

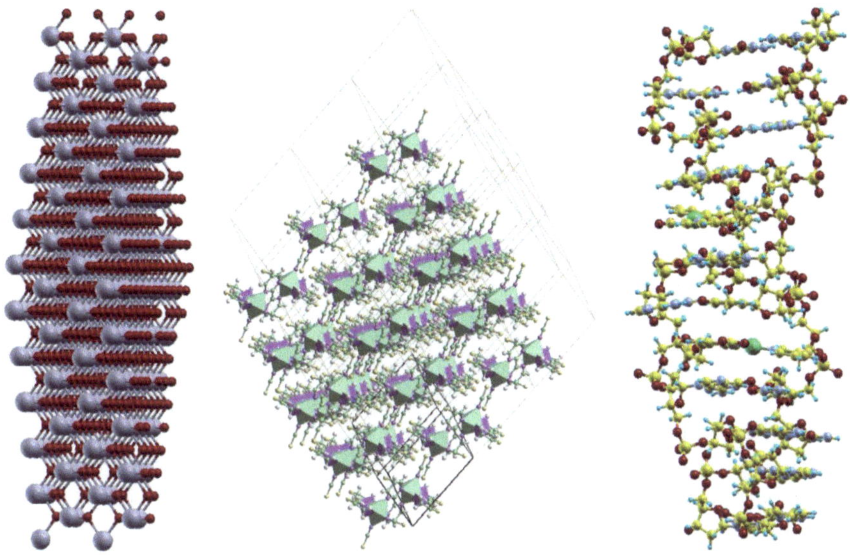

**Fig. 1.1** Some large strongly-correlated systems of interest, in order of decreasing density of correlated sites: a slab of the catalytic multiple-valency lanthanoid oxide $CeO_2$ with a partially reduced surface (*left*); the crystal structure of the canonical binuclear molecular magnet $[\{Fe(bt)(NCS)_2\}_2$ bpym] (bt=2,2-bithiazoline, bpym=2,2-bipyrimidine) [15] (*middle*); and a modified DNA structure [16] with artificial $Cu^{2+}$ mediated pyridine−2−,6−dicarboxylate : pyridine base pairs (*right*)

used to project out the ground-state Kohn–Sham orbitals, namely Wannier functions of both the orthogonal and generalised nonorthogonal varieties. We also address the study of large strongly-correlated systems with linear-scaling cost, in particular those too large to study with conventional cubic-scaling approaches.

In this chapter, we offer a broad introduction to ab initio simulation with Kohn–Sham DFT, with a particular emphasis on linear-scaling methods. The foundations of the theory are first detailed, as well as its limitations. The devices used in practical application of the theory are discussed, including the Born–Oppenheimer, exchange-correlation functional, periodic boundary condition and pseudopotential approximations and their ranges of applicability. We go on to describe the formulation of Kohn–Sham DFT in terms of the single-particle density matrix, the attenuation of the spatial non-locality of which is used, for systems in which it is appropriate, as an artifice by which the linear-scaling of computational effort with system size may be achieved.

In Chap. 2, we describe the physics of strongly-correlated materials and molecules, explaining the failure of conventional local-density type exchange-correlation functionals for such systems and motivating the construction of localised corrections for self interaction to alleviate these problems. We offer an introduction to a popular and effective example of such a correction, the DFT + $U$ method, and give the details of our linear-scaling implementation of this technique in the ONETEP code. We demonstrate its power in scaling tests on nano-clusters of the canonical strongly-correlated insulator nickel oxide.

Chapter 3 addresses the question of optimising the spatial form of the Hubbard projectors used to delineate the strongly correlated subspaces used in locally-corrected ab initio methods such as DFT + $U$. We investigate the projector-dependence of the ground-state properties of a prototypical correlated molecular system, iron porphyrin, and, in order to alleviate this ambiguity in the method with respect to the choice of Hubbard projectors, we propose the use of generalised Wannier functions as Hubbard projectors in a novel self-consistency scheme. We demonstrate the performance of this projector self consistent DFT + $U$ on the same molecular system. This chapter has been published in abbreviated form in Ref. [17].

We continue the theme of optimising subspace projections in Chap. 4, published in Ref. [18], where we consider the optimal way in which to construct a projection operator from a set of nonorthogonal spanning vectors, namely Hubbard projectors. By maintaining the tensorial invariance of all scalars computed using such a projection, as well as a number of other physically-motivated properties, we derive a solution to this problem which differs to the numerous alternatives available in the literature. We discuss the effect of our formalism, with comparison to the most comprehensive one previously proposed, in a projector self-consistent DFT + $U$ study of both bulk nickel oxide and the copper phthalocyanine dimer.

In the closing chapters of the dissertation, we present an original study of the geometric and tensorial consequences of optimising a nonorthogonal representation, which bears fruit in the form of beneficial and hitherto unused corrective terms which, for example, preserve the density-matrix to first order when the representation is updated. This also furnishes a new approach for optimising models for strong correlations. Chapter 5 describes original research on the geometric aspects of using a set of nonorthogonal support functions, or representation vectors, for the single-particle density matrix. In particular, observations concerning the Riemannian manifold associated with the support functions offer new insights into the behaviour of support function optimisation and how this may be made more efficient. Geometric considerations allow us to solve an important problem in linear-scaling ab initio methods, that is a rigorous theory for density-matrix idempotency preservation under changes to the support functions. Chapter 6 provides details of a numerical study on conjugated polymer molecules of an implementation of the most immediately pertinent results of this theory in a modern linear-scaling DFT code.

In Chap. 7, finally, we discuss the challenge of computing the Hubbard interaction tensor $U$, generalising two methods which have been previously proposed to the case of nonorthogonal Hubbard projectors, with linear-scaling expense. We make use of the geometric arguments of Chap. 5 to investigate how this tensor should be changed when the Hubbard projectors are optimised, opening the way for efficient DFT + $U$ calculations for large systems which are self-consistent over both the Hubbard projectors and interaction tensors. We conclude, in Chap. 8, by summarising the main results of the dissertation and suggesting some future avenues for research.

## 1.3 The Born–Oppenheimer Approximation

We now introduce the background to the methodological developments introduced later in the dissertation. We first put the basic principles behind the practical utilisation of Kohn–Sham DFT on a firm footing and then focus on topics relating to its linear-scaling implementation. In practice, the construction of a contemporary linear-scaling DFT code and the introduction of any new functionality, such as described in Chaps. 2, 3 and 6, involves a substantial investment in software development and optimisation for parallel-computing architectures. We do not attempt to detail matters of computer science in this dissertation but we emphasise physical concepts. We begin by providing some of those pre-requisite to linear-scaling DFT in this chapter.

We are generally concerned with the interacting quantum mechanical system composed of $N$ electrons and $\mathcal{N}$ atomic nuclei, located at positions $\mathbf{r}_i$ and $\mathbf{R}_I$, respectively, and described by the Hamiltonian

$$\hat{H} = -\frac{1}{2}\left[\sum_{i=1}^{N} \nabla^2_{\mathbf{r}_i} + \sum_{I=1}^{\mathcal{N}} \frac{1}{M_I}\nabla^2_{\mathbf{R}_I}\right] + \frac{1}{2}\sum_{i=1}^{N}\sum_{j\neq i}^{N}\frac{1}{|\mathbf{r}_i - \mathbf{r}_j|}$$
$$-\sum_{I=1}^{\mathcal{N}}\sum_{i=1}^{N}\frac{Z_I}{|\mathbf{R}_I - \mathbf{r}_i|} + \frac{1}{2}\sum_{I=1}^{\mathcal{N}}\sum_{J\neq I}^{\mathcal{N}}\frac{Z_I Z_J}{|\mathbf{R}_I - \mathbf{R}_J|}. \tag{1.1}$$

Here $\{M_I\}$ and $\{Z_I\}$ are the masses and charges of the nuclei and Hartree atomic units are used, so that $\hbar = e = m_e = 4\pi\epsilon_0$.

For simplicity, we neglect the possibility of an externally-applied electromagnetic field, effects resulting from the finite volume of the nuclei and assume that we are in the non-relativistic regime. Since the Hamiltonian is time-independent, its solution may be written as a product of purely spatial and purely time-dependent functions, as in

$$\Psi(\{\mathbf{r}_i\}, \{\mathbf{R}_I\}) = \Phi(\{\mathbf{r}_i\}, \{\mathbf{R}_I\})\,\Theta(t). \tag{1.2}$$

The energy eigenvalues $\varepsilon$ of the electronic-ionic, or *vibronic*, system are given by the solution of the time-independent Schrödinger equation [19], that is given by

$$\hat{H}|\Phi(\{\mathbf{r}_i\}, \{\mathbf{R}_I\})\rangle = \varepsilon|\Phi(\{\mathbf{r}_i\}, \{\mathbf{R}_I\})\rangle, \tag{1.3}$$

where $|\Phi\rangle$ is the time-independent many-body wave-function. The time-dependent part $|\Theta\rangle$ then evolves, for a given eigenvalue and initial phase $\theta_0$, according to the time-dependent Schrödinger equation, so that

$$i\frac{\partial}{\partial t}\Theta(t) = \varepsilon\Theta(t) \quad \Rightarrow \quad \Theta(t) = \exp(i\theta_0 - i\varepsilon t). \tag{1.4}$$

In principle, $|\Psi\rangle$ completely describes all physical properties of the combined electronic and ionic system. In practice, however, we must make some approximations, beginning with the Born–Oppenheimer approximation described in this section, in order to render the computation of these properties tractable.

Assuming that the vibronic wave-function may be constructed using functions which depend on the electron-nuclear distances, $\mathbf{r}_i - \mathbf{R}_I$, it is clear that the expectation values of its Laplacian with respect electronic and ionic positions are identical. Since, however, the mass of the typical nucleus is a number of orders of magnitude greater than that of the electron, the average electron velocity is expected to be several orders of magnitude greater than that of the nuclei. The motion of the electrons is thus effectively instantaneous with respect to the time scale of nuclear motion, meaning that the electrons may relax to their lowest-energy configuration very quickly for a given ionic configuration. The Born–Oppenheimer adiabatic approximation [20] is invoked to separate the electronic and ionic degrees of freedom, treating the electrons as though they were moving in a static potential due to the nuclei, by assuming a separable form for the vibronic wave-function, so that it is given by

$$\Phi\left(\mathbf{r}_i, \mathbf{R}_I\right) = \Psi\left(\mathbf{r}_i; \mathbf{R}_I\right) \eta\left(\mathbf{R}_I\right). \tag{1.5}$$

An electronic Schrödinger equation may then be defined in terms of a given set of nuclear positions, $\{\mathbf{R}_I\}$, so that

$$\sum_{i=1}^{N}\left[-\frac{1}{2}\nabla_{\mathbf{r}_i}^2 + \frac{1}{2}\sum_{j\neq i}\frac{1}{|\mathbf{r}_i - \mathbf{r}_j|} - \sum_{I=1}^{\mathcal{N}}\frac{Z_I}{|\mathbf{R}_I - \mathbf{r}_i|}\right]\Psi = E\Psi, \tag{1.6}$$

the solution of which we will hereafter be primarily concerned.

Operating with the full Hamiltonian of Eq. 1.1 on the separable vibronic wave-function, of Eq. 1.5, allows, in principle, for solution of the motion of the atomic nuclei. The eigenvalues and eigenfunctions of the electronic Schrödinger equation, Eq. 1.6, depend parametrically on the nuclear positions, however, and the second component of the Born–Oppenheimer approximation is the neglect of vibronic coupling terms involving the gradient and Laplacian of the electronic wave-function with respect to nuclear positions. The solutions of the resulting ionic Schrödinger equation, given by

$$\sum_{J=1}^{\mathcal{N}}\left[-\frac{1}{2M_J}\nabla_{\mathbf{R}_J}^2 + E\left(\{\mathbf{R}_I\}\right) + \frac{1}{2}\sum_{J,K\neq J}^{\mathcal{N}}\frac{Z_J Z_K}{|\mathbf{R}_J - \mathbf{R}_K|}\right]\eta = \varepsilon\eta, \tag{1.7}$$

provide the nuclear dynamics within this approximation, where the adiabatic potential energy surface $E\left(\{\mathbf{R}_I\}\right)$ is given by the position dependent equilibrium energy of the electronic system.

In many physical systems, the Born–Oppenheimer approximation is quite adequate to describe the properties of interest. Furthermore, in practice, due to the substantial masses of the nuclei and consequent particulate nature, it is often sufficient to solve for the motion of the nuclei, using the classical analogue of Eq. 1.7, via molecular dynamics methods [13, 21]. The neglect of the vibronic coupling terms is not appropriate for all systems and phenomena, however, and by precisely constraining the nuclei to the adiabatic surface we cannot describe any superconductivity, vibrational spectroscopy, photochemistry or fast-ion effects such as radiation damage.

## 1.4 Density Functional Theory

Numerous techniques, approximate or effectively exact, have been developed to solve the many-body Schrödinger equation of Eq. 1.6. It is currently quite computationally impossible, however, to routinely apply many-body methods to systems composed of the number of atoms, i.e., hundreds to hundreds of thousands, required to capture the complexity of most systems of interest. In general, ignoring for the moment the spin degrees of freedom and assuming a real-valued many-particle $N$-electron wave-function $\Psi$ discretised on grid of $M$ points in each of three Cartesian directions, $\Psi$ is a function of $3N$ coordinates and so $M^{3N}$ scalars are needed to describe it [14]. For realistic atomistic simulations, these numbers become intractably large, and so the many-particle wave-function is truly beyond reach. Moreover, for many observables of interest, it is not straightforward to construct expectation values from the many-body wave-function, even if it were known. In any case, it is only necessary to find the momentum-dependent density of states, the electron density and the density pair-correlation function for a theoretical description of most physical observables [22].

In their seminal work of 1964, Hohenberg and Kohn (HK) rigorously showed, in two theorems, that all ground-state properties of an interacting many-electron system may be expressed, at least in principle, as a functional of its ground-state electron density distribution [23] and, crucially for practical application of the theory, that the ground-state may be located by variational minimisation of the energy with respect to the density. Returning to the numerical example given above, the electron density is described by $M^3$ scalars and it is thus a much simpler and more tractable function than the wave-function. The formalism in question, in which the electron density is promoted as the quantity central to the description of the ground-state properties of many-electron systems [1], is the acclaimed and highly successful *density functional theory* (DFT) .

We begin our introduction to DFT by defining the electron density $n(\mathbf{r})$ for an $N$-electron system described by a normalised ($\langle\Psi|\Psi\rangle = N$) quantum state vector $|\Psi\rangle$, assumed to be antisymmetric under particle exchange so as to satisfy the Pauli exclusion principle [24], by

$$n(\mathbf{r}) = \langle\Psi|\hat{n}|\Psi\rangle = \int \prod_{i=2}^{N} d\mathbf{r}_i \; |\Psi(\mathbf{r}, \mathbf{r}_2, \cdots, \mathbf{r}_N)|^2. \tag{1.8}$$

We denote, for a given static local external potential, $V(\mathbf{r})$, including all system-specific details such as the ionic potentials, the ground-state wave-function by $|\Psi_0\rangle$.

We may decompose the electronic Hamiltonian as $\hat{H}_{el} = \hat{F} + \hat{V}$, where $\hat{F}$ is the universal (possessing an identical form for all systems) contribution consisting of the kinetic energy and electron-electron interaction terms, namely

$$\hat{F} = \hat{T} + \hat{V}_{int} = \sum_{i=1}^{N} -\frac{1}{2}\nabla^2_{\mathbf{r}_i} + \frac{1}{2}\sum_{j\neq i}^{N} \frac{1}{|\mathbf{r}_i - \mathbf{r}_j|}. \tag{1.9}$$

For a given number of electrons, the external potential fully determines the Hamiltonian and hence all of the resulting physical observables. The eigenstates of the Hamiltonian may be found by solving Eq. 1.6; and the energy $E$ of a given state $|\Psi\rangle$ is given by its expectation value with the Hamiltonian,

$$E = \langle\Psi|\hat{H}_{el}|\Psi\rangle = \langle\Psi|\hat{F}|\Psi\rangle + \int d\mathbf{r}\; V_{ext}(\mathbf{r})\, n(\mathbf{r}). \tag{1.10}$$

In particular, the ground-state, $|\Psi_0\rangle$, and thus its charge density, denoted $n_0(\mathbf{r})$, are dependent only on the external potential. In their seminal work, Hohenberg and Kohn [23] showed that the converse is also true, that the mapping between the space of all external potentials and the space of non-degenerate v-representable ground-state densities (the class of v-representable densities are those which are derived from solving the Schrödinger equation with some external potential) is also injective, so that:

**First HK theorem** There exists a one-to-one mapping between the ground-state density $n_0(\mathbf{r})$ of an $N$-electron system and the external potential $V(\mathbf{r})$ acting upon it.

We may prove this statement by contradiction. Suppose that there is another external potential $V'(\mathbf{r})$, differing from $V(\mathbf{r})$ by more than an additive constant, but which gives rise to the same ground-state density $n_0(\mathbf{r})$. Let us denote the ground-states of the Hamiltonians constructed with $V(\mathbf{r})$ and $V'(\mathbf{r})$ by $|\Psi_0\rangle$ and $|\Psi_0'\rangle$, respectively, with corresponding ground-state energies $E^0$ and $E'^0$. Then, using Eq. 1.10, we may write

$$\begin{aligned} E^0 &< \langle\Psi'|\hat{H}_{el}|\Psi'\rangle = \langle\Psi'|\hat{H}'_{el}|\Psi'\rangle + \langle\Psi'|\hat{H}_{el} - \hat{H}'_{el}|\Psi'\rangle \\ &= E'^0 + \int d\mathbf{r}\; \left[V(\mathbf{r}) - V'(\mathbf{r})\right] n(\mathbf{r}) \end{aligned} \tag{1.11}$$

and, simply swapping the states and potentials, conversely,

$$\begin{aligned} E'^0 &< \langle\Psi|\hat{H}'_{el}|\Psi\rangle = \langle\Psi|\hat{H}_{el}|\Psi\rangle + \langle\Psi|\hat{H}'_{el} - \hat{H}_{el}|\Psi\rangle \\ &= E^0 + \int d\mathbf{r}\; \left[V'(\mathbf{r}) - V(\mathbf{r})\right] n(\mathbf{r}). \end{aligned} \tag{1.12}$$

Addition of the inequalities of Eqs. 1.11 and 1.12 yields the contradiction $E^0 + E'^0 < E^0 + E'^0$, invalidating our assumption that two non-trivially different external potentials may give rise to the same ground-state density and proving the theorem. Thus, remarkably, the ground-state wave-function and hence the ground-state expectation value of any observable $\hat{O}$ may, in principle, be calculated as a functional only of $n_0(\mathbf{r})$, as in

$$O_0 = O[n_0] = \langle \Psi_0[n_0] | \hat{O} | \Psi_0[n_0] \rangle. \tag{1.13}$$

The constrained-search approach to DFT [25, 26], extends the HK formalism to all densities that can be obtained from a (possibly degenerate) antisymmetric $N$-electron wave-function (that class known as N-representable) and we now follow this method to reproduce another important result, the second Hohenberg–Kohn theorem.

We first define, by searching over the set of $N$-body antisymmetric wave-functions which give rise to the ground-state density $n(\mathbf{r})$, the universal functional $F[n]$ as the minimum possible expectation value of the internal-energy operator $\hat{F}$ with that density. Thus, we may write

$$F[n] = \min_{n[\Psi]\rightarrow n} \langle \Psi | \hat{F} | \Psi \rangle = \langle \Psi_{[n]} | \hat{F} | \Psi_{[n]} \rangle, \tag{1.14}$$

where $|\Psi_{[n]}\rangle$ is the state which is found to reproduce $n(\mathbf{r})$ and to minimise the internal energy.

Using this, we may uniquely define the lowest energy of the density $n(\mathbf{r})$, when immersed in an arbitrary potential $V(\mathbf{r})$, as

$$E_V[n] = F[n] + \int d\mathbf{r}\, V(\mathbf{r})\, n(\mathbf{r}) = \langle \Psi_{[n]} | \hat{F} + \hat{V} | \Psi_{[n]} \rangle \tag{1.15}$$

Given this external potential, there exists a ground-state wave function $|\Psi_0\rangle$ which provides the lowest possible energy $E_V^0$, as well as reproducing the ground-state density $n_0(\mathbf{r})$, so that by the variational principle, for all N-representable $n(\mathbf{r})$,

$$E_V^0 \leq E_V[n]. \tag{1.16}$$

Noting that $|\Psi_0\rangle$ generates the ground-state density $n_0(\mathbf{r})$, however, we find that

$$\begin{aligned} \min_{n[\Psi]\rightarrow n_0} \langle \Psi | \hat{F} | \Psi \rangle &\leq \langle \Psi_0 | \hat{F} | \Psi_0 \rangle \Rightarrow \\ \min_{n[\Psi]\rightarrow n_0} \langle \Psi | \hat{F} | \Psi \rangle + \int d\mathbf{r}\, n_0(\mathbf{r}) V(\mathbf{r}) &\leq \langle \Psi_0 | \hat{F} | \Psi_0 \rangle + \langle \Psi_0 | \hat{V} | \Psi_0 \rangle \Rightarrow \\ E_V[n_0] &\leq E_V^0. \end{aligned} \tag{1.17}$$

Since $E_V^0$ is, by definition, the ground-state energy, and combining expressions 1.16 and 1.17, we are led to the statement of the second Hohenberg–Kohn theorem,

**Second HK theorem** For all N-representable densities $n(\mathbf{r})$, $E_V[n_0] \leq E_V[n]$, where $E_V^0$ is the ground-state energy of an $N$-electron system, with ground-state density $n_0(\mathbf{r})$, corresponding to the external potential $V(\mathbf{r})$.

The Hohenberg–Kohn theorems demonstrate that the task of calculating the ground-state energy of an interacting $N$-electron system in the presence of a fixed external potential is equivalent to the variational problem of minimising the energy functional $E_V[n]$ with respect to the density. The minimum of $E_V[n]$ is attained at the ground-state density and other extrema of this functional are attained at densities corresponding to excited states of the system, although these are not guaranteed to reproduce the full spectrum of electronic excitations [27]. If the ground state is degenerate, even in the case that some of the degenerate ground-state wave-functions produce the same density, the energy functional remains well defined [28].

Unfortunately, however, no exact method exists to carry out the variational minimisation procedure described above. In particular, no universally-reliable approximation for the universal functional of Eq. 1.14, for which the Hohenberg–Kohn theorem guarantees existence and which must include all many-body interaction effects, exists [14]. The Kohn–Sham approach, which recasts DFT as a single-particle theory via an effective potential and which we will now describe, is the most widely used in practical applications of the theory.

## 1.5 The Kohn–Sham Equations

Conceptually, the Kohn–Sham approach to DFT, proposed in Ref. [29], can trace its origin to soon after the introduction of the Schrödinger equation, when Hartree introduced his self-consistent field method for approximately solving for atomic wave-functions and energies [30]. This method involves iteratively solving the Schrödinger equation for a single particle, subject to an approximate, self-consistently updated, effective potential due to the atomic nuclei and other electrons. The required anti-symmetry of the wave-function, hence the exchange repulsion between electrons, was neglected, a shortcoming which was later rectified by Fock; see Ref. [31, 32]. The latter method, known as "Hartree-Fock", remains a valuable tool for quantum chemistry calculations in spite of neglecting electronic correlation effects. Approximating, in effect, the functional $F[n]$ by a sum of the kinetic, $T_{TF}[n]$, and electrostatic, $U_H[n]$, energies of a classical charge distribution $n(\mathbf{r})$, the Thomas-Fermi approximation [33, 34] may be viewed as another early precursor to Kohn–Sham DFT [22], one in which

$$\begin{aligned} F[n] &\approx T_{TF}[n] + U_H[n] \\ &\equiv \frac{3}{10}\left(3\pi^2\right)^{\frac{2}{3}} \int d\mathbf{r}\; n^{\frac{5}{3}}(\mathbf{r}) + \frac{1}{2}\iint d\mathbf{r}d\mathbf{r}'\; \frac{n(\mathbf{r})\,n(\mathbf{r}')}{|\mathbf{r}-\mathbf{r}'|}. \end{aligned} \tag{1.18}$$

Simple classical approximations such as this do not lead to qualitatively accurate results for anything but the simplest atoms, however, for instance predicting molecules to be unstable in some cases [35].

The approach proposed by Kohn and Sham, in Ref. [29], was to construct an exact mapping between the interacting $N$-electron system and a reference system of $N$ non-interacting fermions. The Hohenberg–Kohn theorems also apply to the reference system but with a greatly simplified internal energy, namely the kinetic energy of $N$ non-interacting fermions, denoted $T_s[n]$. Remarkably, Kohn and Sham demonstrated that there generally exists an effective potential, denoted $V_{KS}(\mathbf{r})$, in the presence of which the ground-state density of the reference system is identical to that of the true interacting system.

In the Kohn–Sham approach, the total-energy functional of the interacting system is first decomposed into constituent contributions as

$$E[n] \equiv T_s[n] + U_H[n] + E_{xc}[n] + \int d\mathbf{r}\, n(\mathbf{r})\, V_{ext}(\mathbf{r}), \tag{1.19}$$

where the exchange-correlation energy $E_{xc}[n]$ is defined to be that due to all many-body quantum effects which are not included in the other terms. Variational minimisation of the energy functional is performed, subject to the particle-number conservation constraint, via

$$\delta\left[E[n] - \mu\left(\int d\mathbf{r}\, n(\mathbf{r}) - N\right)\right] = 0, \tag{1.20}$$

giving the Euler–Lagrange equation in terms of the internal-energy functional $F[n]$, external potential and chemical potential $\mu$,

$$\frac{\delta F[n]}{\delta n(\mathbf{r})} + V_{ext}(\mathbf{r}) = \mu. \tag{1.21}$$

Since the non-interacting reference system is required to also obtain its minimum at the same density, variational minimisation of the Kohn–Sham energy must yield an equivalent expression, and specifically we may write this in the decomposed form

$$\frac{\delta T_s[n]}{\delta n(\mathbf{r})} + \int d\mathbf{r}' \frac{n(\mathbf{r}')}{|\mathbf{r} - \mathbf{r}'|} + \frac{\delta E_{xc}[n]}{\delta n(\mathbf{r})} + V_{ext}(\mathbf{r}) = \mu. \tag{1.22}$$

Next, defining, uniquely by justification of the first Hohenberg–Kohn theorem, an effective Kohn–Sham potential as

$$V_{KS}[n](\mathbf{r}) = \int d\mathbf{r}' \frac{n(\mathbf{r}')}{|\mathbf{r} - \mathbf{r}'|} + V_{xc}[n](\mathbf{r}) + V_{ext}(\mathbf{r}), \tag{1.23}$$

with the exchange-correlation potential given by

$$V_{xc}[n](\mathbf{r}) = \frac{\delta E_{xc}[n]}{\delta n(\mathbf{r})}, \tag{1.24}$$

we see that Eq. 1.22 may be cast into the form of Eq. 1.21. Therefore, the Euler–Lagrange equation of the non-interacting reference system with potential $\hat{V}_{KS}$,

$$\frac{\delta T_s[n]}{\delta n(\mathbf{r})} + V_{KS}[n](\mathbf{r}) = \mu, \tag{1.25}$$

must yield identically the same ground-state density as the original interacting $N$-electron system.

Since the kinetic energy of the non-interacting reference system may be expressed as a sum of its contributions from a set of single-particle orbitals, $\psi_i(r)$, specifically

$$T_s[n] = -\frac{1}{2}\sum_{i=1}^{N} f_i \int d\mathbf{r}\, \psi_i^*(\mathbf{r})\, \nabla^2 \psi_i(\mathbf{r}), \tag{1.26}$$

where $f_i$ is the occupation number of the $i^{th}$ orbital, it is natural to solve Eq. 1.25 using a set of $N$ single-particle Schrödinger equations,

$$\hat{H}_{KS}\psi_i = \left[-\frac{1}{2}\nabla^2 + \hat{V}_{KS}(\mathbf{r})\right]\psi_i = \epsilon_i \psi_i. \tag{1.27}$$

These are the Kohn–Sham equations and their non-interacting single-particle eigenfunctions, named Kohn–Sham orbitals, reproduce the ground-state density of the interacting system via

$$n(\mathbf{r}) = \sum_{i=1}^{N} f_i |\psi_i(\mathbf{r})|^2. \tag{1.28}$$

The independent-particle energy of the reference system, sometimes known as the band energy, is given by the sum of the eigenvalues of the Kohn–Sham system

$$E^{IP} = \sum_{i=1}^{N} f_i \epsilon_i = T_s[n] + \int d\mathbf{r}\, V_{KS}(\mathbf{r})\, n(\mathbf{r}), \tag{1.29}$$

and it may be corrected to give that of the original interacting system by adding corrective terms, as per

$$E = E^{IP} - U_H[n] - \int d\mathbf{r}\, V_{xc}(\mathbf{r})\, n(\mathbf{r}) + E_{xc}[n]. \tag{1.30}$$

The Kohn–Sham problem, due to its non-linearity in the density, must be iteratively solved by alternately optimising the density and updating the Kohn–Sham potential accordingly, at each iteration, until a self-consistent solution is reached [1, 13].

The Kohn–Sham DFT formalism is, if the exact functional $E_{xc}[n]$ is known, a rigorous theory for the ground-state properties of an electronic system; one which greatly simplifies the task of solving for the ground-state density. In many cases, furthermore, the Kohn–Sham eigenvalues, and corresponding orbitals, may provide a good approximate description of the true spectrum of electronic excitations and related spectroscopic properties. However, any such interpretation of the individual solutions to the Kohn–Sham equations must be performed with care since they may potentially differ greatly from the true excitation spectrum of the Hamiltonian in some materials, more usually when many-body correlation effects are substantial [36].

## 1.6 Exchange, Correlation and the Local Density Approximation

Inevitably, since it encapsulates all complex many-particle interactions in a functional of a scalar field, the explicit form of the exchange-correlation $E_{xc}[n]$ is unknown for all but the simplest model systems.

By definition, $E_{xc}[n]$ is composed of the difference between the interacting and noninteracting kinetic energies, given by $T_c[n] = T[n] - T_s[n]$, and the difference between the interacting and electrostatic Coulomb energies, written as $U_c[n] = U[n] - U_H[n]$. Following Ref. [14], it is instructive to separate $E_{xc}[n]$ into a sum of terms purely due to exchange, $E_x[n]$, and correlation effects, $E_c[n]$.

A density-functional form of the exchange energy $E_x[n]$ is not known, although Fock showed that it may be exactly expressed in terms of single-particle orbitals generating the density [32], explicitly

$$E_x = -\frac{1}{2}\sum_{i,j=1}^{N} f_i f_j \iint d\mathbf{r}\, d\mathbf{r}' \frac{\psi_i^*(\mathbf{r})\,\psi_j^*(\mathbf{r}')\,\psi_i(\mathbf{r}')\,\psi_j(\mathbf{r})}{|\mathbf{r}-\mathbf{r}'|}. \tag{1.31}$$

The exchange effect is due to the Pauli exclusion principle, the fact that no two fermions may occupy the same quantum state [24], and it lowers the Coulomb repulsion energy by spatially separating electrons of equal spin. When using a wave-function in the form of a single Slater determinant, the single-particle kinetic energy $T_s[n]$ is exactly described by Eq. 1.26 and so the the kinetic energy correction $T_c[n]$ is a part of $E_c[n]$, that is purely a correlation effect [3, 14].

The correlation energy, $E_c[n]$, in turn, may be interpreted as the difference between the true ground-state energy and that obtainable with a single self-consistent Slater determinant. The correlation effect includes all many-body repulsion between electrons of opposite spin and it always lowers the total-energy. There is no universally accepted definition of a "strongly correlated system", although it is clear that this is a reasonable description if the magnitude of $E_c[n]$ is comparable to that of $T_s[n]$ or $U_H[n]$ [14]. For such systems, some of which are of great interest and which we describe in detail in Chap. 2, we may be justifiably doubtful of the accuracy of the single-determinant Kohn–Sham wave-function computed using mean-field approximations to $E_{xc}[n]$.

The exchange-correlation energy may be recast in the form of a classical electrostatic interaction between the charge depletion hole density, $n_{xc}(\mathbf{r}, \mathbf{r}')$, due to the effects of exchange and correlation, and the inhomogeneous electron density distribution $n(\mathbf{r})$ inducing it [37, 38], using the expression

$$E_{xc}[n] = \frac{1}{2}\iint d\mathbf{r}\, d\mathbf{r}' \frac{n(\mathbf{r})n_{xc}(\mathbf{r}, \mathbf{r}')}{|\mathbf{r}-\mathbf{r}'|}, \tag{1.32}$$

where $n_{xc}(\mathbf{r}, \mathbf{r}')$ may be defined in terms of the density pair-correlation function.

Importantly, it may be shown that the exact $E_{xc}[n]$ depends only on the spherical average of $n_{xc}(\mathbf{r}, \mathbf{r}')$ and that it also satisfies the sum rule [39] given, where $n_x(\mathbf{r}, \mathbf{r}')$ is the charge depletion generated by exchange only, by

$$\int d\mathbf{r}' \, n_{xc}(\mathbf{r}, \mathbf{r}') = \int d\mathbf{r}' \, n_x(\mathbf{r}, \mathbf{r}') = -1. \tag{1.33}$$

The Local Density Approximation (LDA) for exchange and correlation appeared early in the development of DFT, and it still maintains much popularity and applicability. In the LDA, the contribution to $E_{xc}[n]$ due to each infinitesimal volume element $d\mathbf{r}$ is approximated by that corresponding to a homogenous electron gas with the same density $n(\mathbf{r})$ in that volume element. We may therefore write

$$E_{xc}^{LDA}[n] = \int d\mathbf{r} \, \epsilon_{xc}^{LDA}(n(\mathbf{r}))\, n(\mathbf{r}) \Rightarrow \tag{1.34}$$

$$V_{xc}^{LDA}(\mathbf{r}) = \frac{\delta E_{xc}^{LDA}[n]}{\delta n(\mathbf{r})} = \epsilon_{xc}^{LDA}(n(\mathbf{r})) + n(\mathbf{r}) \left. \frac{d\epsilon_{xc}^{LDA}(n)}{dn} \right|_{n=n(\mathbf{r})},$$

where $\epsilon_{xc}^{LDA}(n(\mathbf{r}))$ is the exchange-correlation energy per electron of a homogeneous electron gas of density $n(\mathbf{r})$. A commonly used LDA functional for practical DFT calculations is the Perdew–Zunger parameterization [40]. This satisfies some known analytical properties of $E_{xc}[n]$, fits the numerical Monte Carlo data of Ceperley and Alder [41] and the analytical high-density expression of Gell-Mann and Brueckner [42].

It has been shown that the LDA reproduces the physical attributes of certain systems which have very inhomogeneous charge densities with reasonable accuracy, perhaps owing to the fact both that it satisfies the sum rule for $n_{xc}(\mathbf{r}, \mathbf{r}')$, Eq. 1.33, and that the spherical averages of the exact $n_{xc}(\mathbf{r}, \mathbf{r}')$ and that predicted by the LDA are often quite similar in form [39]. However, the LDA exhibits the tendency (partly due to spurious self-interaction effects, e.g., the self-evident conditions $E_c[n] = 0$ and $E_x[n] = -U_H[n]$ are not satisfied by the LDA applied to a one-electron system [14]) towards systematic overestimation of cohesive binding energies, consequently underestimated bond lengths, underestimated local moments, and underestimated insulating energy gaps (the ionisation potential and electron affinity are reproduced by the exact functional).

Numerous advances upon the LDA have been developed in order to improve the description of the complex electronic interactions occurring in real systems. More sophisticated approximations to the exchange-correlation functional include, for example, hybrid functionals which involve some fraction of the exact exchange [36], the optimised effective potential method [43], self-interaction [39] and orbital-polarisation corrections [44] to the LDA and the DFT+$U$ method [45, 46] which we introduce in Chap. 2. No method has yet been developed which systematically and faithfully reproduces the physics of a large range of systems while incurring a favourable computational expense. Generalised Gradient Approximations (GGA), however, are semi-local extensions to the LDA which include information on spatial fluctuations in the density, e.g.,

$$E_{xc}^{GGA}[n] = \int d\mathbf{r}\, \zeta\left(n(\mathbf{r}), |\nabla n(\mathbf{r})|, \nabla^2 n(\mathbf{r}), \cdots\right) \tag{1.35}$$

and, particularly in molecular systems, often offer some improvements over the LDA. The widely-used GGA functional of Perdew, Burke and Ernzerhof (PBE) is constructed using analytical conditions known for the exact functional and it provides satisfactory results in a wide variety of systems [47]. It is the exchange-correlation functional predominantly employed in this dissertation.

## 1.7 Spin-Density Functional Theory

We have hitherto treated the total-energy as a functional of the charge density alone, including the spin-dependent nature of the Pauli exclusion principle in the difference between exchange and correlation effects but otherwise ignoring the degrees of freedom introduced by the electron spin. This formalism is inadequate for many systems, including all open-shell molecules and extended systems that exhibit spontaneous magnetisation. Here we restrict ourselves to the case of collinear electron spins, however, although the theory of DFT for non-collinear magnetism has been developed to describe canted magnetisation which, for example, is characteristic of systems, such as those comprising rare-earth metals, where spin-orbit coupling plays an important role [48]. It is then sufficient to treat the density of spin-up and spin-down electrons, written as $n^{\uparrow}(\mathbf{r})$ and $n^{\downarrow}(\mathbf{r})$, respectively, separately in order to provide the system the variational freedom to spontaneously form magnetic structures or to interact to an externally-applied magnetic field (at least in the Zeeman limit).

In this widely-used approach, known as spin DFT (SDFT) [38, 49], individual Kohn–Sham equations are solved for the up and down-spin components of the density, labelled by the spin index $\sigma$, so that

$$\hat{H}_{KS}^{\sigma}\psi_i^{(\sigma)} = \left[-\frac{1}{2}\nabla^2 + V_{KS}^{(\sigma)}(\mathbf{r})\right]\psi_i^{(\sigma)} = \epsilon_i^{(\sigma)}\psi_i^{(\sigma)}. \tag{1.36}$$

The total spin density $n(\mathbf{r})$, the magnetisation density $m(\mathbf{r})$ and the internal magnetic field, $B_{xc}(\mathbf{r})$, may be calculated, respectively, using

$$n(\mathbf{r}) = n^{\uparrow}(\mathbf{r}) + n^{\downarrow}(\mathbf{r}), \tag{1.37}$$

$$m(\mathbf{r}) = \mu_0\left(n^{\uparrow}(\mathbf{r}) - n^{\downarrow}(\mathbf{r})\right) \quad \text{and} \tag{1.38}$$

$$B_{xc}(\mathbf{r}) = \mu_0 \left( V_{xc}^{\uparrow}(\mathbf{r}) - V_{xc}^{\downarrow}(\mathbf{r}) \right), \tag{1.39}$$

where $\mu_0$ is the Bohr magneton [14]. The Hartree term remains unchanged in the SDFT formalism and so the spin-dependence of the potential enters only through $\hat{V}_{xc}^{(\sigma)}$. The analogue of LDA for SDFT, dubbed LSDA [49], is usually constructed using interpolation procedures which preserve certain properties such as the theoretically exact spin-scaling relation for exchange [50], given by

$$E_x^{SDFT}\left[n^{\uparrow}, n^{\downarrow}\right] = \frac{1}{2}\left(E_x^{DFT}\left[2n^{\uparrow}\right] + E_x^{DFT}\left[2n^{\downarrow}\right]\right). \tag{1.40}$$

## 1.8 The Pseudopotential Approximation

We now describe an additional computational expedient, the well-known *pseudopotential* approximation [51–53], for obviating the explicit treatment of tightly-bound core-level atomic orbitals which are observed to participate little in, and to be only slightly perturbed by, chemical bonding. The valence eigenstates, whose energy-levels are much more susceptible to the atomic environment due to the substantial cancellation between the ionic Coulomb potential and the repulsive screening effects of the core electrons, are retained for evaluation. The stipulation of orthogonality between the valence and the more spatially localised core electrons necessitates rapid spatial oscillations in the former and, consequently, a prohibitively great number of basis functions, or Fourier components, are needed to describe them accurately. The computational economy afforded by the pseudopotential approximation thus greatly exceeds that simply due to the reduction of the number of eigenstates to be solved for. Moreover, with care, the removal of the core electrons may improve the accuracy of total-energy differences under changes to the external potential, since these will comprise a significantly greater proportion of the total energy.

Following Ref. [54], let us consider a valence state $|\psi^{valence}\rangle$ which is an eigenvalue of an atomic Hamiltonian $\hat{H}$ with eigenvalue $E$. We suppose that we may construct a smoother pseudo-state $|\psi^{pseudo}\rangle$ by subtracting a linear-combination of core states, so that

$$|\psi^{valence}\rangle = |\psi^{pseudo}\rangle + \sum_{n}^{core} a_n |\chi_n\rangle. \tag{1.41}$$

The required orthogonality between the valence state and each core state actually fixes the coefficients $a_n$, and, with Eq. 1.41, this gives

$$0 = \langle \chi_n | \psi^{valence} \rangle = \langle \chi_n | \psi^{pseudo} \rangle + a_n \Rightarrow$$
$$a_n = -\langle \chi_n | \psi^{pseudo} \rangle$$
$$|\psi^{valence}\rangle = |\psi^{pseudo}\rangle - \sum_n^{core} |\chi_n\rangle\langle\chi_n|\psi^{pseudo}\rangle. \tag{1.42}$$

This expression may be immediately re-cast into a more useful eigenvalue equation for the pseudised state, namely

$$\left[ \hat{H} + \sum_n^{core} (E - E_n) |\chi_n\rangle\langle\chi_n| \right] |\psi^{pseudo}\rangle = E|\psi^{pseudo}\rangle, \tag{1.43}$$

so that the correct valence eigenvalue $E$ is reproduced if the pseudo-Hamiltonian differs from the original original by the potential

$$\hat{V}^{non-local}(E) = \sum_n^{core} (E - E_n) |\chi_n\rangle\langle\chi_n|. \tag{1.44}$$

The sum of the ionic potential and $\hat{V}^{non-local}(E)$, which is repulsive and localised to the core-region, provides a shallow smoothly-varying pseudopotential for the valence electrons.

The core states $\{|\chi_n\rangle\}$ and their energy levels $E_n$, for each atomic species, are usually fixed throughout, having been first computed in an all-electron isolated-atom DFT calculation for an appropriate electronic configuration. The energy argument $E$ of the pseudopotential is, in practice, evaluated at the corresponding valence-state energy of the atomic system; the validity of this approximation, known as the *transferability* of the pseudopotential, depends on the extent to which, for all core levels $n$ and where $\Delta E$ is the change to $E$ in moving from the atomic to chemical environment, $E - E_n \gg \Delta E$ holds.

In order to properly preserve the scattering properties of the pseudised core states, a different pseudopotential must be computed for valence states of different angular momentum $l$, although independently of the azimuthal quantum number $m$ for spherically-symmetric atomic potentials. Supposing that the potential due to core electrons vanishes beyond a cutoff radius $r_c$, the valence wave-functions are expressible as the separable product of a radial function $R_l$ and spherical harmonic $Y_{lm}$,

$$\psi_{lm}(\mathbf{r}; E) = R_l(\mathbf{r}; E)\, Y_{lm}(\theta, \phi). \tag{1.45}$$

It may be shown that the scattering phase shift $\delta_l$ of a plane-wave of energy $E = \frac{1}{2}k^2$ is given by

$$\left. \frac{d}{dr} log\left[R_l(\mathbf{r}; E)\right] \right|_{r=r_c} = k \frac{j_l'(kr_c) - tan(\delta_l)\, n_l'(kr_c)}{j_l(kr_c) - tan(\delta_l)\, n_l(kr_c)}, \tag{1.46}$$

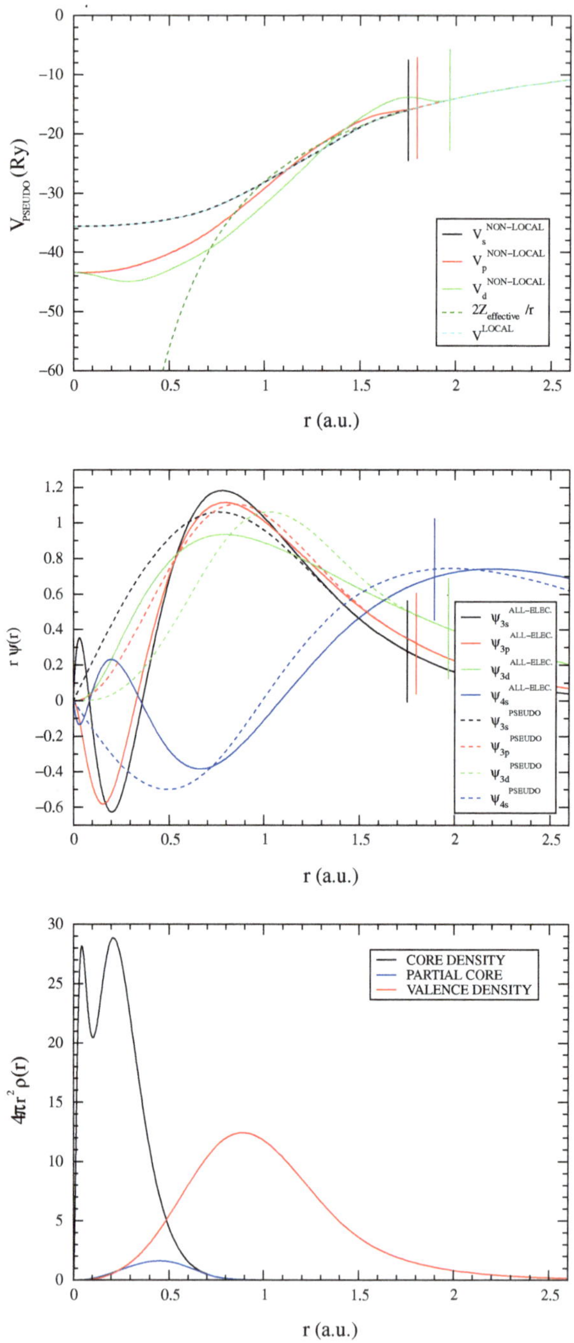


**Fig. 1.2** Pseudopotential for atomic chromium (A set of RRKJ Pseudopotentials were generated using Opium http://pium.sourceforge.net, and the GGA input parameters available therein, optimized for a minimum plane-wave cutoff of 680 eV, albeit with a scalar-relativistic correction for all species and, for the transition-metal ions, some slight modifications to the core radii and a non-linear core correction of Fuchs–Scheffler characteristic radius 1.3a.u.) [55, 56]. Ionic and pseudised potentials (*top panel*); the $l = 0$ channel is chosen as local. Probability amplitude of the all-electron and pseudised atomic wave-functions, including semi-core 3*s* electrons as valence states (*middle panel*). Density distribution of core, valence and partial-core ($r_{NLCC} = 0.77$ a$_0$) manifolds (*bottom panel*)

where $j_l(kr)$ and $n_l(kr)$ are the spherical Bessel and spherical von Neumann functions, respectively. Furthermore, for a given cutoff radius and radial function evaluated there, $R_l(\mathbf{r}_c; E)$, the energy derivative of this expression is proportional to the norm of the wave-function in the core region; the latter is given by

$$4\pi \int_0^{r_c} dr\, r^2 R_l^2(r). \tag{1.47}$$

As a result, by forcing $R_l^{pseudo}$ to satisfy both Eqs. 1.46 and 1.47 through stipulating that, for each required angular momentum channel,

$$R_l^{pseudo}(r) = R_l^{valence}(r) \quad \text{for all } r \geq r_c, \tag{1.48}$$

we may construct *norm-conserving* pseudopotentials; we refer the reader to Refs. [55, 57–59] for technical details, which preserve the scattering properties of the original atomic potential up to first order in *E*. A locally-acting pseudopotential is usually also included to represent the average effects of angular momenta higher than the maximum represented, $l_{max}$, so that generally we have

$$\hat{V}^{pseudo} = \hat{V}^{local} + \sum_l^{l_{max}} \sum_{m=-l}^{l} |Y_{lm}\rangle \hat{V}_l^{non-local} \langle Y_{lm}|. \tag{1.49}$$

An important caveat in applying the pseudopotential approximation to spin polarised atomic species where there is a significant spatial overlap between densities due to core and valence electrons, respectively $n_c$ and $n_v$, notably in first-row transition-metal ions onwards, is that, unlike the Hartree contribution, the exchange-correlation potential is generally linear in neither the charge density nor magnetisation density. In particular, it is usually assumed in pseudopotential construction that the valence contribution, $n_v$, to the pseudopotential exchange-correlation contribution

$$V_{xc}^{(\sigma)}[n_c + n_v, \xi] - V_{xc}^{(\sigma)}[n_v, \xi_v] \tag{1.50}$$

completely cancels. It does not in general, but the approximation may be particularly hazardous when the pseudised and all-electrons spin polarisations, respectively defined as

$$\xi_v(\mathbf{r}) = \frac{n_v^{(\uparrow)}(\mathbf{r}) - n_v^{(\downarrow)}(\mathbf{r})}{n_v(\mathbf{r})} \quad \text{and} \quad \xi(\mathbf{r}) = \frac{n_v^{(\uparrow)}(\mathbf{r}) - n_v^{(\downarrow)}(\mathbf{r})}{n_v(\mathbf{r}) + n_c(\mathbf{r})}, \tag{1.51}$$

significantly differ [60]. The non-linear core correction (NLCC) is a first-order method by which a smoothly-varying component of the core spin-density, the so-called partial core, with characteristic smoothing radius $r_{NLCC}$, is retained for addition to the valence density upon computation of the exchange-correlation potential and energy [56, 60]. A numerical example of a norm-conserving, non-local, non-linear core corrected pseudopotential is shown in Fig. 1.2.

## 1.9 Periodicity and Brillouin Zone Sampling

In this section and the next, we begin to discuss some of the technical details of the linear-scaling DFT method, known as ONETEP [61, 62], which we primarily employ for our numerical studies. Even with the simplifications to the task of solving for the ground-state density discussed in previous sections, the explicit treatment of the Kohn–Sham orbitals in extended systems such as solids, molecular crystals etc., which may have effectively infinite spatial extent as well as infinite cardinality remains, with no further artifice, computationally intractable.

However, if we may reasonably suppose that the atomic nuclei in the system at hand are arranged in a perfectly periodic lattice, with primitive cell vectors $\{\mathbf{a}_j\}$, then the (non-relativistic) Kohn–Sham potential also has the periodicity of the Bravais lattice, so that

$$V_{KS}(\mathbf{r}+\mathbf{R}) = V_{KS}(\mathbf{r}); \quad \mathbf{R} = \sum_{i=j}^{3} n_j \mathbf{a}_j; \quad n_i \in \mathbb{Z}. \tag{1.52}$$

In this case, we may make use of Bloch's theorem, which we briefly demonstrate below, to reformulate our stated problem into one for the solution of a finite number of Kohn–Sham eigenequations, labelled by a continuous set of crystal momentum vectors **k** in the first Brillouin zone. For a more detailed treatment of the principles outlined in this section see Refs. [63, 64].

We begin by noting that since the periodicity of the potential implies the periodicity of the Hamiltonian, then, with translation operators $\hat{T}_{\mathbf{x}}$ defined to advance the vector argument of scalar functions by a displacement **x**, we may write that

$$\begin{aligned}
\hat{T}_{\mathbf{R}} H(\mathbf{r}) &= H(\mathbf{r}+\mathbf{R}) = H(\mathbf{r}) \Rightarrow \\
\hat{T}_{\mathbf{R}}\left[H(\mathbf{r})\,\psi_i(\mathbf{r})\right] &= H(\mathbf{r}+\mathbf{R})\,\psi_i(\mathbf{r}+\mathbf{R}) = H(\mathbf{r})\,\hat{T}_{\mathbf{R}}\psi_i(\mathbf{r}) \Rightarrow \\
\left[\hat{H}, \hat{T}_{\mathbf{R}}\right] &= 0, \text{ for all } \mathbf{R} \text{ given by Eq. 1.52}, \Rightarrow \\
\hat{T}_{\mathbf{R}}\psi_{\mathbf{R}}(\mathbf{r}) &= c(\mathbf{R})\,\psi_{\mathbf{R}}(\mathbf{r}), \text{ where } c \in \mathbb{C}.
\end{aligned} \tag{1.53}$$

Here we have used the fact that the commutativity of the Hamiltonian and lattice translation implies that quantum numbers simultaneously exist to label eigenfunctions of both operators.

The exponentiation identity $c\left(\mathbf{R}+\mathbf{R}'\right) = c(\mathbf{R})\,c\left(\mathbf{R}'\right)$ immediately suggests a natural representation for the periodic eigenvalues of lattice translation, specifically

$$c(\mathbf{R}) = \exp(\iota \mathbf{k}\cdot\mathbf{R}); \quad \mathbf{k} = \sum_{j=1}^{3} x_j \mathbf{b}_j; \quad \mathbf{a}_l \cdot \mathbf{b}_j = 2\pi\delta_{lj}, \tag{1.54}$$

where the complex numbers $\{x_j\}$ are general pre-multipliers for the reciprocal lattice vectors $\{\mathbf{b}_j\}$. It is clear from the above that crystal momentum vectors, **k** and $\mathbf{k}'$, which differ by a reciprocal lattice vector, $\mathbf{k} - \mathbf{k}' = \mathbf{G} = \sum_{j=1}^{3} n_j \mathbf{b}_j$, are equivalent for our purposes.

Next, considering the function $u_{i\mathbf{k}}(\mathbf{r}) = \exp(-\iota \mathbf{k} \cdot \mathbf{r}) \psi_{i\mathbf{k}}(\mathbf{r})$, for a given eigenstate of the Hamiltonian, labelled $i$, we find that

$$\begin{aligned} \hat{T}_{\mathbf{R}} u_{i\mathbf{k}}(\mathbf{r}) &= \exp(\iota \mathbf{k} \cdot (\mathbf{r} - \mathbf{R})) \psi_{i\mathbf{k}}(\mathbf{r} + \mathbf{R}) \\ &= \exp(-\iota \mathbf{k} \cdot \mathbf{r}) \psi_{i\mathbf{k}}(\mathbf{r}) = u_{i\mathbf{k}}(\mathbf{r}). \end{aligned} \tag{1.55}$$

Thus, we conclude that the eigenfunctions $\{\psi_{i\mathbf{k}}\}$ of the spatially-periodic Hamiltonian $\hat{H}$ may be written (the statement known as Bloch's theorem) as

$$\psi_{i\mathbf{k}}(\mathbf{r}) = \exp(\iota \mathbf{k} \cdot \mathbf{r}) u_{i\mathbf{k}}(\mathbf{r}), \tag{1.56}$$

i.e., as the product of a cell-periodic function $u_{i\mathbf{k}}(\mathbf{r})$ and a plane-wave of complex wave-vector $\mathbf{k}$. The reformulation of the Kohn–Sham equations into Bloch form appears as a set of Hermitian eigenvalue equations for the cell-periodic functions $u_{i\mathbf{k}}(\mathbf{r})$, where a dependence on the Bloch vector $\mathbf{k}$ enters the the Hamiltonian operator and, consequently, the band occupancies $f_{i\mathbf{k}}$.

We suppose, next, that we may neglect any effects due to the boundary of the system under study and that we may therefore impose periodic (Born–von Kármán) boundary conditions to our sample, let us say of $N_{sample} = N_1 \times N_2 \times N_3$unit cells, so that

$$\psi_{i\mathbf{k}}(\mathbf{r}) = \psi_{i\mathbf{k}}\left(\mathbf{r} + N_j \mathbf{a}_j\right). \tag{1.57}$$

As an immediate result, the set of permitted Bloch vectors $\mathbf{k}$ is reduced to a discrete set of $N_{sample}$ real-valued vectors which, as we have shown, may be restricted to the first Brillouin zone since they are equivalent modulo any reciprocal lattice vector $\mathbf{G}$.

For infinite systems, where $N_{sample} \to \infty$, the number of distinct $\mathbf{k}$-points in the Brillouin zone diverges and, in principle, since the occupied Kohn–Sham states at each point contribute to the electronic potential for the periodic sample [13], continuous integration over the Brillouin zone is required in order to calculate the electronic density and Hamiltonian. However, since the eigenstates of the Hamiltonian vary smoothly between proximate $\mathbf{k}$-points [65], it may be shown that Brillouin zone integrals may be well-approximated by sampling at a small number of carefully selected and weighted Bloch vectors [66, 67]. The magnitude of the error thereby introduced is reduced by increasing the finesse of $\mathbf{k}$-point sampling, whereupon the total-energy converges, not necessarily monotonically, to its limiting value.

Bloch's theorem does not strictly apply, of course, to systems where translational symmetry is broken in one or more directions, such as molecular systems, defective crystals and layered heterostructures. In principle, a complex-valued continuous reciprocal-space sampling must be reintroduced in the direction in which the symmetry is broken. This technical difficulty is usually circumvented by using a periodically-repeated supercell filled with a sufficient quantity of the medium surrounding the object under scrutiny (e.g., vacuum surrounding a molecule, or pristine crystal surrounding a point defect) to ensure that it is isolated (both in the sense of orbital overlap and of electrostatic interactions) from its artificial periodic images [68]. In practical calculations, the supercell volume is increased until the total energy is adequately converged.

For large systems of interest for linear-scaling methods, e.g., for the supercell volumes $\Omega_{cell}$ necessary to study molecular systems or defective crystals, under periodic boundary conditions, the first Brillouin zone volume

$$\Omega_{1^{st}BZ} = \frac{(2\pi)^3}{\Omega_{cell}} \tag{1.58}$$

is sufficiently small that the **k**-dependence of the Kohn–Sham eigenstates is negligible. In this regime, a single point, usually the Γ-point (zone centre $\mathbf{k} = 0$), is used to sample reciprocal space (the eigenstates may then be assumed real-valued). In order to emulate a calculation which uses a regular grid of $N_k$ vectors in each reciprocal-lattice directions using a Γ-point only formalism, an example of which is described in Chap. 4, an explicit repetition of the supercell by a factor of $N_k$ in each Bravais lattice direction is used to provide an equivalent sampling of the Bloch momentum dependence.

## 1.10 The Plane-Wave and Psinc Basis Sets

It is necessary, in practice, to restrict the Hilbert space available to the supercell-periodic Kohn–Sham wave-functions to a computationally manageable subspace spanned by a finite set of *basis functions*. A good choice of basis set provides for efficient evaluation of spatially-dependent operators, such as potentials, and differential operators, which are usually computed in reciprocal space. Moreover, it is advantageous if the basis can be systematically improved using variational parameters or if the basis functions reflect some of the physical nature of atomic systems, e.g., form eigenvectors of angular momentum and possess a suitable number of radial nodes, so that a small set may efficiently reproduce the ground-state density.

A truncated basis set of plane-waves is very common choice for periodic DFT calculations, particularly in conjunction with the pseudopotential approximation previously described, since it allows for very efficient Fourier transformation using FFT techniques. The supercell-periodic part of the **k**-dependent Kohn–Sham wave-functions may be expanded as a Fourier series of discrete plane-waves, labelled by the wave-vectors **G** of the reciprocal lattice, so that the functions themselves admit a plane-wave representation given by

$$\begin{aligned} \psi_{i\mathbf{k}}(\mathbf{r}) &= e^{\iota\mathbf{k}\cdot\mathbf{r}} u_{i\mathbf{k}}(\mathbf{r}) = e^{\iota\mathbf{k}\cdot\mathbf{r}} \left[ \sum_{\mathbf{G}} c_{i\mathbf{k},\mathbf{G}} e^{\iota\mathbf{G}\cdot\mathbf{r}} \right] \\ &= \sum_{\mathbf{G}} c_{i,\mathbf{k}+\mathbf{G}} e^{\iota(\mathbf{k}+\mathbf{G})\cdot\mathbf{r}}. \end{aligned} \tag{1.59}$$

While, in principle, an infinite set of plane-waves is needed to fully reproduce the wave-functions at each **k**-point, a truncated Fourier expansion may be used in practice since the contributions beyond a system-dependent bandwidth are negligible. A kinetic cut-off energy $E_{cut}$ is used as the single variational parameter in plane-wave methods; meaning that wave-vectors satisfying $\frac{1}{2}|\mathbf{k}+\mathbf{G}|^2 \leq E_{cut}$ are retained in the wave-function expansion and the resulting error in the total-energy is monotonically reduced by increasing $E_{cut}$, augmenting the basis, until acceptable convergence is obtained [13].

In DFT methods for which the computational effort scales linearly with system size, as we go on to discuss, it is necessary to attenuate the interactions between spatially well-separated areas of the supercell. This may be carried out using efficient localised (approximately atomic-like) basis functions or, as in the case of the ONETEP method [61, 62] on which we concentrate hereon, the concatenation of a variational basis set with such a set of functions. The latter, known as the *representation, support functions* [69], project out the space of Kohn–Sham orbitals and are afforded the variational freedom of the basis within certain approximations, as we later detail. In order to retain the systematic improvability and orthogonality of a plane-wave basis, but with the spatial localisation required to efficiently expand spatially-localised support functions $\{\phi_\alpha(\mathbf{r})\}$, the elegant approach used in ONETEP is to use the real-space representation of a truncated set of plane-waves.

These, variously known as Fourier-Lagrange, periodic bandwidth limited delta, or *psinc* functions [70, 71], are defined by

$$D_{\{m\}}(\mathbf{r}) = \prod_{i=1}^{3} \frac{1}{N_i} \sum_{p_i=-J_i}^{J_i} e^{\iota (p_i \mathbf{b}_i) \cdot (\mathbf{r}-\mathbf{r}_{\{m\}})}, \tag{1.60}$$

where the number of grid-points in each lattice direction is $N_i = 2J_i + 1,\ J_i \in \mathbb{N}$, and are immediately recognisable as approximations to Dirac delta functions at the points $\mathbf{r}_m$, since

$$D_{\{m\}}(\mathbf{r}) \approx \delta(\mathbf{r} - \mathbf{r}_m) = \frac{\Omega_{cell}}{(2\pi)^3} \int d\mathbf{G}\, e^{\iota \mathbf{G} \cdot (\mathbf{r}-\mathbf{r}_{\{m\}})}. \tag{1.61}$$

The *psinc* basis is thus wholly equivalent to a plane-wave basis truncated to a rhombohedron of reciprocal lattice vectors and the finesse of the real-space grid, represented by

$$\mathbf{r}_{\{m\}} = \sum_{i=3}^{3} \frac{m_i}{N_i} \mathbf{a}_i; \quad m_i \in \{0, 1, \cdots, N_i - 1\}, \tag{1.62}$$

must be increased until the energy and other properties of interest are well converged. The stated kinetic energy cutoff of a ONETEP calculation is approximate, however, since it is computed using the reciprocal-space radius generating a sphere with the same volume of the aforementioned rhombohedron.

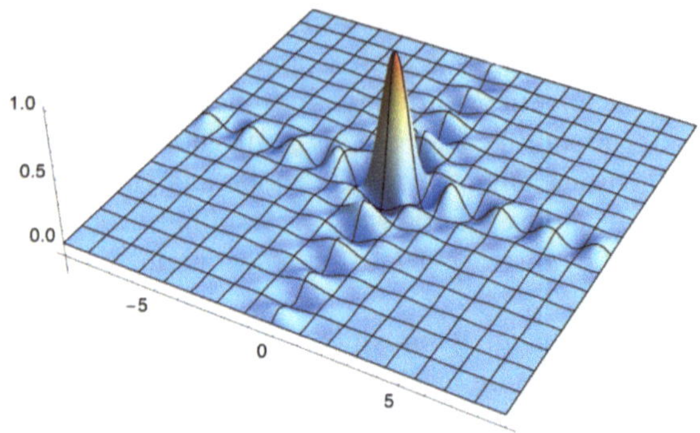


**Fig. 1.3** Periodic cardinal sine, or *psinc*, function centred at the origin of a two-dimensional square lattice of periodicity $N_{i,j} = 21$. The function has unit value at its centre-point and zero value at all inequivalent grid points

The cell-periodicity $D_{\{m\}}(\mathbf{r}) = D_{\{m\}}(\mathbf{r}+\mathbf{R})$ of the *psinc* functions may be shown, as may the fact that each function is localised to one grid-point alone and well-normalised, so that $D_{m_1 m_2 m_3}\left(\mathbf{r}_{\mu_1 \mu_2 \mu_3}\right) = \delta_{m_1 \mu_1}\delta_{m_2 \mu_2}\delta_{m_3 \mu_3}$ and real-valued everywhere in the supercell [70]. The graph of a particular *psinc* function, which limits to a cardinal sine function as $N_i \rightarrow \infty$, is depicted in Fig. 1.3. The evaluation of the overlap integrals between functions represented on the *psinc* grid and the matrix elements of local and differential operators are all calculable using efficient Fourier transforms and parallelised sums over grid points. The overlap integral of a real-valued cell-periodic support function $\phi_\alpha(\mathbf{r})$ and a $\mathbf{r}_{\{m\}}$-centred *psinc function*, denoting $N_{cell} = N_1 \times N_2 \times N_3$,

$$\int d\mathbf{r}\ D_{\{m\}}(\mathbf{r})\,\phi_\alpha(\mathbf{r}) = \frac{1}{N_{cell}} \sum_{\mathbf{G}}^{\mathbf{G}_{max}} \phi_\alpha(\mathbf{G})\, e^{\iota \mathbf{G}\cdot\mathbf{r}_{\{m\}}} = \frac{\Omega_{cell}}{N_{cell}} \phi_\alpha^D\left(\mathbf{r}_{\{m\}}\right), \tag{1.63}$$

is exactly the *psinc*-centred evaluation of the bandwidth-limited approximation to $\phi_\alpha$ [70]. The orthogonality of the *psincs* follows as a very convenient consequence of this, which we can show by replacing $\phi$ by any $D_{\{m'\}\neq\{m\}}$. The value of a support function at an arbitrary location $\mathbf{r}$ may, conversely, be represented as a summation over the simulation cell via

$$\phi_\alpha^D(\mathbf{r}) = \sum_{\{m\}} C_{\{m\},\alpha} D_{\{m\}}(\mathbf{r}); \quad \text{where} \tag{1.64}$$

$$C_{\{m\},\alpha} = \phi_\alpha^D\left(\mathbf{r}_{\{m\}}\right) \quad \text{and} \quad \sum_{\{m\}} \equiv \sum_{q_1=0}^{N_1-1} \sum_{q_2=0}^{N_2-1} \sum_{q_2=0}^{N_3-1}.$$

In practice, all spatial functions are approximated by their Fourier-filtered counterparts and so, hereafter, we suppress the corresponding superscript $D$.

## 1.11 Density-Matrix Formulation of DFT

Kohn–Sham DFT methods in which the eigenstates are explicitly expanded in the basis typically incur a cubic asymptotic scaling of computational effort with respect to the number of atoms, $\mathcal{N}$, in the supercell; a prohibitive cost for $\mathcal{N} > 10^3$ at present. The orthonormality condition

$$\delta_{ij} = \int\limits_{\Omega_{cell} \propto \mathcal{N}} d\mathbf{r}\ \psi_i^* (\mathbf{r})\, \psi_j (\mathbf{r})\,; \quad \forall\, i, j \in \{1, \cdots, N\}\,; \quad N \propto \mathcal{N}, \tag{1.65}$$

which must be maintained during energy minimisation, makes the spatial non-locality of the eigenstates abundantly clear to be the fundamental cause of this computational bottleneck. Much progress has been made in recent years with DFT methods which realise the intrinsic linear-scaling nature of the density-functional and we refer the reader to Refs. [72, 73] for comprehensive reviews. Here, in particular, we focus on methods, such as those described in Refs. [61, 69, 74–78], which explicitly minimise the total-energy functional with respect to the density-matrix representation of the zero-temperature Kohn–Sham wave-function in order to circumvent the explicit imposition of Eq. 1.65. In doing so, as we explain in the following section, we may exploit the "near-sighted" nature of the quantum mechanics [79].

The density-matrix of a quantum system at temperature $T$ is generally given by the normalised Hermitian partition function,

$$\hat{\gamma} = \frac{e^{-\beta \hat{H}}}{Tr[e^{-\beta \hat{H}}]} = \frac{\sum_i e^{-\beta E_i} |\Psi_i\rangle\langle\Psi_i|}{\sum_j e^{-\beta E_j}}, \quad \text{where} \quad \beta = \frac{1}{k_B T}, \tag{1.66}$$

from which we may evaluate the statistical average of a many-body operator $\hat{O}$ via the trace

$$O = \langle \hat{O} \rangle = Tr\left[\hat{O}\hat{\gamma}\right] = \frac{\sum_i e^{-\beta E_i} \langle\Psi_i|\hat{O}|\Psi_i\rangle}{\sum_j e^{-\beta E_j}}. \tag{1.67}$$

Usually, in the context of Kohn–Sham DFT, we are concerned only with the zero temperature limit

$$\hat{\gamma} = \sum_i |\Psi_i\rangle\langle\Psi_i|. \tag{1.68}$$

Of rather more practical utility, however, is the integral of $\hat{\gamma}$ over $N-1$ and $N-2$ co-ordinates (for $N$-particle systems), which yields the reduced single-particle and two-particle density matrices, $\rho\left(\mathbf{r}, \mathbf{r}'\right)$ and $\rho_2\left(\mathbf{r}_1, \mathbf{r}_2; \mathbf{r}'_1, \mathbf{r}'_2\right)$, respectively. The single-particle $\hat{\rho}$ is sufficient to evaluate the expectation values of single-particle operators (e.g., the frequency-integrated Green's function $\hat{\rho}\left(\mathbf{r}, \mathbf{r}'\right) = -\imath \lim_{t' \to t} \hat{G}\left(t - t'\right)$, local potentials and kinetic energy) and $\hat{\rho}_2$ is needed for two-particle operators (e.g., the many-body Hamiltonian, two-particle Coulomb interactions, the exchange-correlation hole) [14]. Significantly, only the diagonal part of $\hat{\rho}$ is needed to compute the density, so that $n(\mathbf{r}) = \rho(\mathbf{r}, \mathbf{r})$ and we may always express the density-functional in terms of the density-matrix. For a periodic system, the wave-function takes the form of a single Slater determinant of single-particle Bloch orbitals $\psi_{i\mathbf{k}}^{(\sigma)}(\mathbf{r})$, with occupancies $f_{i\mathbf{k}}^{(\sigma)}$, so that

$$\langle \mathbf{r}|\hat{\rho}^{(\sigma)}|\mathbf{r}'\rangle = \frac{\Omega_{cell}}{(2\pi)^3} \sum_i \int_{1^{st}BZ} d\mathbf{k}\; f_{i\mathbf{k}}^{(\sigma)} \psi_{i\mathbf{k}}^{(\sigma)}(\mathbf{r})\, \psi_{i\mathbf{k}}^{(\sigma)*}(\mathbf{r}') . \tag{1.69}$$

The density-matrix idempotency,

$$\hat{\rho}^{(\sigma)2} = \hat{\rho}^{(\sigma)} \Rightarrow \rho^{(\sigma)}(\mathbf{r}, \mathbf{r}') = \int d\mathbf{r}''\; \rho^{(\sigma)}(\mathbf{r}, \mathbf{r}'')\, \rho^{(\sigma)}(\mathbf{r}'', \mathbf{r}') , \tag{1.70}$$

is the defining condition for a pure, fermionic, spin-collinear state, and it ensures that $f_{i\mathbf{k}}^{(\sigma)} \in \{0, 1\}$, for all $i$, $\mathbf{k}$ and $\sigma$, and thus that the Pauli exclusion principle [24] is respected. The satisfaction of the idempotency requirement, as well as particle-number conservation, $Tr\left[\hat{\rho}^{(\sigma)}\right] = N^{(\sigma)}$, is the foremost challenge in constructing methods with which to minimise functionals in the density-matrix representation. Numerous strategies have been proposed to drive the density-matrix to idempotency during total-energy minimisation, the longest established of which simply add a idempotency-deviation penalising term $P\left[\hat{\rho}\right]$ to the energy functional, such as the functional of McWeeny [80],

$$\widetilde{E}\left[\hat{\rho}\right] = E\left[\hat{\rho}\right] + \alpha P\left[\hat{\rho}\right]; \quad P\left[\hat{\rho}\right] = Tr\left[\left(\hat{\rho} - \hat{\rho}^2\right)^2\right]. \tag{1.71}$$

The minimum of the functional $\widetilde{E}$ approaches that of $E$, from below, with an error scaling with the inverse of the penalty pre-factor $\alpha$ [81].

The McWeeny functional is employed in ONETEP in a number of different guises; see Ref. [81] for a complete overview. Most simple of these, however, is the purifying transformation $\hat{\rho}_{m+1} = 3\hat{\rho}_m^2 - 2\hat{\rho}_m^3$, equivalent to a recursive steepest descents minimisation of $P\left[\hat{\rho}\right]$ and depicted in Fig. 1.4. For three density matrices, each with sets of eigenvalues $\left\{f_i^1\right\}$, $\left\{f_i^2\right\}$ and $\left\{f_i^3\right\}$ which satisfy, for all $i$,

$$f_i^1 \in \left[\frac{1-\sqrt{5}}{2}, \frac{1+\sqrt{5}}{2}\right]; \; f_i^2 \in \left[\frac{1-\sqrt{3}}{2}, \frac{1+\sqrt{3}}{2}\right]; \; f_i^3 \in \left[\frac{-1}{2}, \frac{3}{2}\right],$$

the purifying transformation is stable; devoid of occupancy flipping about $f_i = 0.5$; or produces a "weakly idempotent" density matrix (i.e., $f_i \in [0, 1]$) in one step, respectively.

When a sufficient level of idempotency has been reached ONETEP using purifying transforms, LNVD method [82–84] may be used to minimise the total energy functional while refining the density-matrix further. In this remarkable technique, the

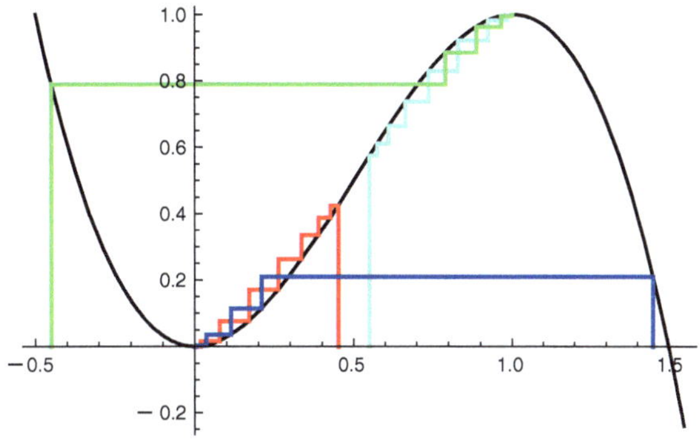


**Fig. 1.4** Orbits of the McWeeny [80] purification transformation $f_{m+1} = 3f_m^2 - 2f_m^3$ for occupancies $f_i$ in the interval $[-0.5, 1.5]$, e.g., $-0.45$ (*green*), 0.45 (*red*), 0.55 (*cyan*) and 1.45 (*blue*)

energy is directly minimised with respect to an auxiliary density-matrix $\hat{\sigma}$, related to the density-matrix by one iteration of the purifying transform, i.e., $\hat{\rho} = 3\hat{\sigma}^2 - 2\hat{\sigma}^3$, for each spin channel. The total-energy is re-expressed in terms of $\hat{\sigma}$ such that $E_{LNVD}\left[\hat{\sigma}\right] = E\left[\hat{\rho}\right]$. Here, however, a generalised Kohn–Sham energy functional, proposed in Ref. [85], is used in order to accommodate non-integer density-matrix eigenvalues $f_i$. The resulting functional $E_{LNVD}$ is minimised only at physically-meaningful integer occupancies, nonetheless, with no multiple minima, meaning that the density-matrix is driven towards idempotency in a computationally efficient manner as the energy is minimised.

A number of modifications to the energy functional have been proposed which provide that the trace of the density-matrix is also driven towards the correct number of electrons, beginning with the LNVD [82] grand potential defined as $\Omega = E - \mu N = Tr[\hat{\rho}(\hat{H} - \mu)]$, where the chemical potential $\mu$ is determined numerically. In ONETEP, the HSMP [81] variant of the LNVD method is used, in which the auxiliary density-matrix is defined in terms of a purified and renormalised density-matrix, so that

$$\hat{\rho} = N \frac{3\hat{\sigma}^2 - 2\hat{\sigma}^3}{Tr\left[3\hat{\sigma}^2 - 2\hat{\sigma}^3\right]}. \tag{1.72}$$

Density-matrix search directions computed using this transformation provide that the gradient of the electron number is projected out of the energy gradient by construction, so that the chemical potential is effectively self-determined.

## 1.12 Wannier Function and Density-Matrix Localisation

A characteristic property of quantum mechanical systems is their "near-sightedness", the fact that locally-evaluated expectation values depend little on the details of spatially-distant parts of the system [79]. A reformulation of the periodic Bloch wave-functions $\psi_{i\mathbf{k}}^{(\sigma)}(\mathbf{r})$ which is suited to the study of localisation properties was proposed by Wannier [86], whose eponymously named functions are defined, and localised to a unit cell at the lattice vector **R**, by

$$|w_{n\mathbf{R}}^{(\sigma)}\rangle = \left(\frac{\Omega_{cell}}{(2\pi)^3}\right)^{\frac{1}{2}} \int\limits_{1^{st}BZ} d\mathbf{k}\; e^{-\imath\mathbf{k}\cdot\mathbf{R}} |\psi_{n\mathbf{k}}^{(\sigma)}\rangle. \tag{1.73}$$

The unitarity of the Fourier transform preserves the orthonormality of the Bloch orbitals, so that

$$\int d\mathbf{r}\; w_{n\mathbf{R}}^{(\sigma)*}(\mathbf{r})\, w_{n'\mathbf{R}'}^{(\sigma)}(\mathbf{r}) = \delta_{nn'}\delta_{\mathbf{R}\mathbf{R}'}. \tag{1.74}$$

Supposing that the band-occupancies are **k**-independent, i.e., restricting ourselves to insulating systems, we may write the single-particle density-matrix in terms of Wannier functions as per

$$\rho^{(\sigma)}\left(\mathbf{r},\mathbf{r}'\right) = \sum_{n}\sum_{\mathbf{R}} w_{n\mathbf{R}}^{(\sigma)}(\mathbf{r})\, f_n^{(\sigma)} w_{n\mathbf{R}}^{(\sigma)*}\left(\mathbf{r}'\right). \tag{1.75}$$

A set of exponentially-localised Wannier functions,

$$w\,(x) \sim \exp\left(-\lambda\,|x - x_0|\right), \quad \text{where } \lambda \propto \sqrt{E_{band-gap}}, \tag{1.76}$$

may be constructed for the one-dimensional tight-binding model [79] and similar exponential decay properties have been rigorously shown also for real insulating crystals [87–89]. This is a crucial result for linear-scaling ab initio methods, since, as a consequence of Eq. 1.75, the elements of the real-space representation of the single-particle density-matrix then also decay exponentially with spatial distance, so that

$$\rho^{(\sigma)}\left(\mathbf{r},\mathbf{r}'\right) \sim \exp\left(-|\mathbf{r}-\mathbf{r}'|\right). \tag{1.77}$$

Thus, we may justifiably truncate the real-space representation of the density-matrix for insulating systems, neglecting elements for which $|\mathbf{r}-\mathbf{r}'| > R_{cut}$, so that the information which needs to be stored scales linearly with system size. The system-dependent characteristic cutoff distance $R_{cut}$ determines the number of atoms beyond which the linear-scaling regime holds, but, since it controls the extent to which the variational freedom of the density-matrix is restricted, it also is a parameter with respect to which convergence must be tested.

It is, of course, computationally impractical to manipulate the expression of the density-matrix on a real-space grid such as that provided by the variational *psinc* basis. A separable expression of $\hat{\rho}$ in terms of a minimal set of spatially localised support functions [69] $\{\phi_i\}$, a special case of which is Eq. 1.75, is employed in practice. In the case of an orthonormal and spin-independent set of support functions, we may write

$$\psi_i^{(\sigma)}(\mathbf{r}) = \sum_{j=1}^{N} \phi_j(\mathbf{r})\, U_{ji}^{(\sigma)}, \tag{1.78}$$

so that

$$\rho^{(\sigma)}\left(\mathbf{r}, \mathbf{r}'\right) = \sum_{ij} \phi_i(\mathbf{r})\, K_{ij}^{(\sigma)} \phi_j^*\left(\mathbf{r}'\right), \tag{1.79}$$

where the *density kernel* [80] is given by

$$K_{ij}^{(\sigma)} = \langle \phi_i | \hat{\rho}^{(\sigma)} | \phi_i \rangle = \sum_n U_{in}^{(\sigma)} f_n^{(\sigma)} U_{nj}^{(\sigma)\dagger}. \tag{1.80}$$

The support functions themselves are usually spatially truncated, in practice, usually to atom-centred spheres of species-dependent radii $r_{cut}$ that are variational parameters. Thus, if support functions $A$ and $B$ are centred on atoms positioned at $\mathbf{R}^A$ and $\mathbf{R}^B$ and cutoff at radii $r_{cut}^A$ and $r_{cut}^B$, respectively, then

$$\left|\mathbf{R}^B - \mathbf{R}^A\right| > R_{cut} + r_{cut}^A + r_{cut}^B \Rightarrow K_{AB}^{(\sigma)} = 0 \tag{1.81}$$

and a linear-scaling increase in the number of non-zero matrix elements of $K^{(\sigma)}$ with system size is achieved.

## 1.13 The ONETEP Method

ONETEP is a linear-scaling total-energy and force code which is based on the principles which we have described in the previous sections. Central to the ONETEP methodology is the utilisation of a set of spatially-truncated support functions in the form of Nonorthogonal Generalised Wannier Functions (NGWFs), suppressing the spin index,

$$|\phi_{\alpha\mathbf{R}}\rangle = \left(\frac{\Omega_{cell}}{(2\pi)^3}\right)^{\frac{1}{2}} \int_{1^{st}BZ} d\mathbf{k}\, e^{-\imath\mathbf{k}\cdot\mathbf{R}} \sum_{i=1}^{N} |\psi_{i\mathbf{k}}\rangle \left[M_i^{\dagger\beta}(\mathbf{k})\, S_{\beta\alpha}\right], \tag{1.82}$$

the expansion coefficients $C_{\{m\},\alpha}$ of which, in a *psinc* basis, are variationally optimised in situ (i.e., not as a pre or post-processing step) in order to minimise the total-energy functional with accuracy equivalent to a plane-wave method [61, 69, 90, 91]. Here, the transformation matrix

$$M_i^{\alpha}(\mathbf{k}) = \int d\mathbf{r}\, \phi^{\alpha}{}_{\mathbf{k}}(\mathbf{r})\, \psi_{i\mathbf{k}}(\mathbf{r}) \tag{1.83}$$

is not necessarily unitary and we make use of the Einstein summation convention proposed in Ref. [92], whereby repeated indices within the same expression are summed over unless the indices are contained in parentheses. We hereafter use Greek indices for nonorthogonal vectors and assume Brillouin zone centre sampling only.

The NGWF overlap matrix is defined and computed as

$$S_{\alpha\beta} \equiv \langle \phi_\alpha | \phi_\beta \rangle = \frac{\Omega_{cell}}{N_{cell}} \sum_{\{m\}} C^*_{\{m\},\alpha} C_{\{m\},\beta} \quad \text{(c.f. Eq. 1.64)}, \tag{1.84}$$

so that we may express the orthonormality of Kohn–Sham orbitals as

$$M_i^{\dagger\alpha} S_{\alpha\beta} M_j^\beta = \delta_{ij}. \tag{1.85}$$

Here, we note that when dealing with nonorthogonal functions, we must distinguish between covariant vectors, e.g., NGWFs $\phi_\alpha(\mathbf{r})$, and the corresponding contravariant vectors, e.g., the NGWF duals $\phi^\beta(\mathbf{r})$, which inter-depend via the bilinear relationship

$$\langle \phi_\alpha | \phi^\beta \rangle = \delta_\alpha^\beta \quad \Rightarrow \quad \phi^\alpha(\mathbf{r}) = \phi_\beta(\mathbf{r})\, S^{\beta\alpha}; \quad S^{\beta\alpha} = \left(S_{\alpha\beta}\right)^{-1}. \tag{1.86}$$

We return to this topic in further detail in Chap. 5. The density kernel in the nonorthogonal case, the generalisation of Eq. 1.80, is conventionally given in terms of the dual functions, so that

$$\langle \phi^\alpha | \hat{\rho} | \phi^\beta \rangle = K^{\alpha\beta} = \sum_i M_n^\alpha f_n M_n^{\dagger\beta}. \tag{1.87}$$

Thus, for example, we may write the total number of electrons and the independent-particle energy, respectively, as

$$N = 2K^{\alpha\beta} S_{\beta\alpha} \quad \text{and} \quad E^{IP} = 2K^{\alpha\beta} \langle \phi_\beta | \hat{H}_{KS} | \phi_\alpha \rangle, \tag{1.88}$$

where the factors of two appear for spin-degeneracy.

An important technical point is that the action of differential operators on the density matrix, such as the Laplacian, for example, are not evaluated in the whole simulation cell, but instead in a smaller rhombohedron of *psinc* functions whose volume is independent of the system size. This expedient, known as the FFTBOX approximation, relies on the smoothness of the NGWFs in reciprocal space due to their real-space localisation, since it effectively reduces the sampling rate of the former. The projection from the simulation cell onto the subspace spanned by the FFTBOX *psinc* functions may be written as an operator $\hat{\mathcal{P}}_{(\beta\alpha)}$, so that we may write the kinetic energy, for example, as

$$T = -K^{\alpha\beta} \langle \phi_\beta | \nabla^2 | \phi_\alpha \rangle \approx -K^{\alpha\beta} \langle \phi_\beta | \hat{\mathcal{P}}^\dagger_{(\beta\alpha)} \nabla^2 \hat{\mathcal{P}}_{(\beta\alpha)} | \phi_\alpha \rangle, \tag{1.89}$$

where the argument $(\beta\alpha)$ locates the FFTBOX at an origin $\mathbf{r}_{(\beta\alpha)}$ such that NGWFs $\phi_\alpha$ and $\phi_\beta$ are fully enclosed. The FFTBOX approximation ensures that the cost for each Fourier transform does not depend on system size, so that the total cost of such operations scales linearly with the number of atoms.

In the ONETEP total energy minimisation scheme, two nested conjugate-gradients constrained search loops are performed in order to minimise the total-energy with respect to the density matrix, in a manner which is reminiscent of that proposed, in a somewhat different context, in Ref. [93]. The NGWFs and the corresponding density kernel which yield the ground-state energy

$$\begin{aligned} E^0 &= \min_n E\left[n\right] = \min_{\hat{\rho}} E\left[\hat{\rho}\right] \\ &= \min_{\hat{\rho}} E\left[\hat{\rho}\left(\left\{K^{\alpha\beta}\right\}, \left\{\phi_\alpha\right\}\right)\right] \\ &= \min_{\left\{K^{\alpha\beta}\right\},\left\{C_{\{m\},\alpha}\right\}} E\left[\left\{K^{\alpha\beta}\right\}, \left\{C_{\{m\},\alpha}\right\}\right] \\ &= \min_{\left\{C_{\{m\},\alpha}\right\}} \mathcal{E}\left[\left\{C_{\{m\},\alpha}\right\}\right]; \quad \text{where} \\ \mathcal{E}\left[\left\{C_{\{m\},\alpha}\right\}\right] &\equiv \min_{\left\{K^{\alpha\beta}\right\}} E\left[\left\{K^{\alpha\beta}\right\}; \left\{C_{\{m\},\alpha}\right\}\right] \end{aligned} \tag{1.90}$$

are located such that the density-matrix is idempotent, well normalised and commutes with the Hamiltonian. A simple schematic of the nested search scheme is shown in Fig. 3.4, where the outer and inner loops of interest to this chapter are coloured in blue and red, respectively.

In the outer energy minimisation loop, the density kernel is kept fixed while the total energy is variationally minimised with respect to the spatial profile of the NGWFs. For an idempotent density-matrix, the outer loop corresponds to solving the Kohn–Sham eigenvalue problem. It should be noted that a number methods have been proposed in which equations of motion are set up for the support functions [78, 94–96], but that the NGWFs in ONETEP are perhaps unusual in that they are afforded the variational freedom of a plane-wave equivalent basis, at least within generous truncation spheres. In the inner loop, the energy is minimised with respect to the density kernel matrix elements, while the NGWF expansion is kept fixed. In this process, we locate the best idempotent density-matrix commuting with its own self-consistent Hamiltonian in the current NGWF representation.

Rather than detail the formulae needed to optimise the NGWFs, density kernel and ionic positions here, we give a detailed example of how to compute the contribution to each, due to a particular correction useful for the study of strongly correlated systems, in Sects. 2.5, 2.4 and 2.6, respectively. For more expansive expositions on the ONETEP method, we refer the reader to Refs. [61, 74, 81, 91, 97].

The techniques that we have described in this chapter amount to a remarkable collective achievement in the field of electronic structure theory, one which provides for a wide variety of systems, including some quite large structures when linear-scaling approaches are used, to be successfully simulated. One of the remaining difficulties, however, is the reliable treatment of strong electron correlation effects, which we have mentioned in Sect. 1.6 and which, in the context of linear-scaling DFT, we now proceed to discuss in more detail.

## References


1. R.M. Martin, *Electronic Structure: Basic Theory and Practical Methods* (Cambridge University Press, New York, 2004), p. 624, (ISBN 0-521-78285-6)
2. R.W. Godby, P.G. González , in *Density Functional Theories and Self-Energy Approaches*, ed. by C. Fiolhais, F. Nogueira, M.A.L. Marques. A Primer in Density Functional Theory, vol. 620, Lecture Notes in Physics (Springer, Heidelberg, 2003)
3. E.K.U. Gross, E. Runge, O. Heinonen, *Many-Particle Theory* (Adam Hilger, Bristol, 1991)
4. L. Hedin, New method for calculating the one-particle green's function with application to the electron-gas problem. Phys. Rev. **139**, A796 (1965)
5. F. Aryasetiawan, O. Gunnarsson, The GW method. Rep. Prog. Phys. **61**, 273 (1998)
6. C. Friedrich, A. Schindlmayr, in *Many-Body Perturbation Theory: The GW Approximation*, ed. by J. Grotendorst, S. Blügel, D. Marx. Computational Nanoscience: Do It Yourself!, vol. 31, NIC Series (John von Neumann Institute for Computing, Jülich, 2006)
7. A. Georges, G. Kotliar, Hubbard model in infinite demensions. Phys. Rev. B **45**, 6479 (1992)
8. G. Kotliar, S.Y. Savrasov, K. Haule, V.S. Oudovenko, O. Parcollet, C.A. Marianetti, Electronic structure calculations with dynamical mean-field theory. Rev. Mod. Phys. **78**, 865 (2006)
9. K. Held, I.A. Nekrasov, G. Keller, V. Eyert, N. Blümer, A. McMahan, R. Scalettar, T. Pruschke, A.I. Anisimov, D. Vollhardt, Realistic investigations of correlated electron systems with LDA+DMFT. Psi-k Newsletter **56**, 65 (2003)
10. G. Kotliar, D. Vollhardt, Strongly correlated materials: insights from dynamical mean-field theory. Phys. Today **57**(3) (2004).
11. E. Runge, E.K.U. Gross, Density-functional theory for time-dependent systems. Phys. Rev. Lett. **52**(12), 997 (1984)
12. P. Elliott, F. Furche, K. Burke, Excited states from time-dependent density functional theory, in *Reviews in Computational Chemistry*, eds. by K.B. Lipkowitz, T.R. Cundari, (Wiley, Hoboken, NJ, 2009), pp. 91–165
13. T.A. Arias, M.C. Payne, J.D. Joannopoulos, Ab initio molecular dynamics techniques extended to large length-scale systems. Phys. Rev. B **45**(4), 1538 (1992)
14. K. Capelle, A bird's-eye view of density-functional theory. Braz. J. Phys **36**, 1318 (2006)
15. A.B. Gaspar, V. Ksenofontov, S. Reiman, P. Gütlich, A.L. Thompson, A.E. Goeta, M.C. Muoz, J.A. Real, Mössbauer investigation of the photoexcited spin states and crystal structure analysis of the spin-crossover dinuclear complex $\{Fe(bt)(NCS)_2\}_2$ bpym. (bt=2,2-Bithiazoline, bpym=2,2-Bipyrimidine). Chem. Eur. J. **12**(36), 9289 (2006)
16. S. Atwell, E. Meggers, G. Spraggon, P.G. Schultz, Structure of a copper-mediated base pair in DNA. J. Am. Chem. Soc. **123**(49), 12364 (2001)
17. D.D. O'Regan, N.D.M. Hine, M.C. Payne, A.A. Mostofi, Projector self-consistent DFT+U using nonorthogonal generalized Wannier functions. Phys. Rev. B **82**(8), 081102 (2010)
18. D.D. O'Regan, M.C. Payne, A.A. Mostofi, Subspace representations in ab initio methods for strongly correlated systems. Phys. Rev. B **83**(24), 245124 (2011)
19. E. Schrödinger, An undulatory theory of the mechanics of atoms and molecules. Phys. Rev. **28**(6), 1049 (1926)
20. M. Born, R. Oppenheimer, Zur quantentheorie der molekeln. Ann. d. Physik **84**(20), 457 (1927)
21. R. Car, M. Parrinello, Unified approach for molecular dynamics and density functional theory. Phys. Rev. Lett. **55**(22), 2471 (1985)
22. V. Antonov, B. Harmon, A. Yaresko, *Electronic Structure and Magneto-Optical Properties of Solids*. (Kluwer Academic/ Dordrecht/ Boston/ London, 2004)

23. P. Hohenberg, W. Kohn, Inhomogeneous electron gas. Phys. Rev. **136**(3B), B864 (1964)
24. W. Pauli, The connection between spin and statistics. Phys. Rev. **58**, 716 (1940)
25. M. Levy, Electron densities in search of Hamiltonians. Phys. Rev. A **26**(3), 1200 (1982)
26. E.H. Lieb, Density functionals for Coulomb-systems. Int. J. Quantum Chem. **24**(3), 243 (1983)
27. J.P. Perdew, M. Levy, Extrema of the density functional for the energy: excited states from the ground-state theory. Phys. Rev. B **31**(10), 6264 (1985)
28. R.M. Dreizler, E.K.U. Gross, *Density Functional Theory, An Approach to the Quantum Many-Body Problem* (Springer, New York, 1990).
29. W. Kohn, L.J. Sham, Self-consistent equations including exchange and correlation effects. Phys. Rev. **140**(4A), A1133 (1965)
30. D.R. Hartree, The wave mechanics of an atom with a non-Coulomb central field. Proc. Camb. Phil. Soc. **24**(1), 89 (1928)
31. J.C. Slater, Note on Hartree's method. Phys. Rev. **35**(2), 210 (1930)
32. V. Fock, Näherungsmethode zur lösung des quantenmechanischen Mehrkörperproblems. Z. Phys. **61**(1–2), 126 (1930)
33. L.H. Thomas, The calculation of atomic fields. Proc. Camb. Phil. Soc. **23**, 542 (1927)
34. E. Fermi, Un metodo statistico per la determinazione di alcune proprietà dell'atome. Rend. Accad. Naz. Lincei **6**, 602 (1927)
35. E. Teller, On the stability of molecules in the Thomas-Fermi theory. Rev. Mod. Phys. **34**(4), 627 (1962)
36. W. Kohn, A. Becke, R. Parr, Density functional theory of electronic structure. J. Phys. Chem. **100**, 12974 (1996)
37. D.C. Langreth, J.P. Perdew, The exchange-correlation energy of a metallic surface. Solid State Commun. **17**(1), 1425 (1975)
38. O. Gunnarsson, B.I. Lundqvist, Exchange and correlation in atoms, molecules, and solids by the spin-density-functional formalism. Phys. Rev. B **13**(10), 4274 (1976)
39. R.O. Jones, O. Gunnarsson, The density functional formalism, its applications and prospects. Rev. Mod. Phys. **61**(3), 689 (1989)
40. J.P. Perdew, A. Zunger, Self-interaction correction to density-functional approximations for many-electron systems. Phys. Rev. B **23**(10), 5048 (1981)
41. D.M. Ceperley, B.J. Alder, Ground state of the electron gas by a stochastic method. Phys. Rev. Lett. **45**(7), 566 (1980)
42. M. Gell-Mann, K.A. Brueckner, Correlation energy of an electron gas at high density. Phys. Rev. **106**(2), 364 (1957)
43. J.B. Krieger, Y. Li, G.J. Iafrate, Construction and application of an accurate local spin-polarized Kohn–Sham potential with integer discontinuity: exchange-only theory. Phys. Rev. A **45**, 101 (1992)
44. O. Eriksson, J. M. Wills, M. Colarieti-Tosti, S. Lebgue, A. Grechnev. Many-body projector orbitals for electronic structure theory of strongly correlated electrons. Int. J. Quantum Chem. **105** (2) (2005)
45. V.I. Anisimov, J. Zaanen, O.K. Andersen, Band theory and Mott insulators: Hubbard U instead of Stoner I. Phys. Rev. B **44**(3), 943 (1991)
46. V.I. Anisimov, I.V. Solovyev, M.A. Korotin, M.T. Czyżyk, G.A. Sawatzky, Density-functional theory and NiO photoemission spectra. Phys. Rev. B **48**(23), 16929 (1993)
47. J.P. Perdew, K. Burke, M. Ernzerhof, Generalized gradient approximation made simple. Phys. Rev. Lett. **77**(18), 3865 (1996)
48. J. Kubler, K.H. Hock, J. Sticht, A.R. Williams, Density functional theory of non-collinear magnetism. J. Phys. F Metal Phys. **18**(3), 469 (1988)
49. von U. Barth, L. Hedin, A local exchange-correlation potential for the spin polarized case. i. J. Phys. C Solid State Phys. **5**(13), 1629 (1972)
50. G.L. Oliver, J.P. Perdew, Spin-density gradient expansion for the kinetic energy. Phys. Rev. A **20**(2), 397 (1979)

51. V. Heine, *The pseudopotential concept*, vol. 24, Solid State Physics (Academic Press, New York, 1970), p. 1ff
52. J.C. Phillips, Energy-band interpolation scheme based on a pseudopotential. Phys. Rev. **112**(3), 685 (1958)
53. J.C. Phillips, L. Kleinman, New method for calculating wave functions in crystals and molecules. Phys. Rev. **116**(2), 287 (1959)
54. C. Herring, A new method for calculating wave functions in crystals. Phys. Rev. **57**(12), 1169 (1940)
55. A.M. Rappe, K.M. Rabe, E. Kaxiras, J.D. Joannopoulos, Optimized pseudopotentials. Phys. Rev. B **41**(2), 1227 (1990)
56. M. Fuchs, M. Scheffler, Ab initio pseudopotentials for electronic structure calculations of polyatomic systems using density-functional theory. Comput. Phys. Commun. **119**(1), 67 (1999)
57. D.R. Hamann, M. Schlüter, C. Chiang, Norm-conserving pseudopotentials. Phys. Rev. Lett **43**(20), 1494 (1979)
58. G.P. Kerker, Non-singular atomic pseudopotentials for solid-state applications. J. Phys. C **13**(9), L189 (1980)
59. D.R. Hamann, Generalized norm-conserving pseudopotentials. Phys. Rev. B **41**(2), 2980 (1989)
60. S.G. Louie, S. Froyen, M.L. Cohen, Nonlinear ionic pseudopotentials in spin-density-functional calculations. Phys. Rev. B **26**(4), 1738 (1982)
61. C.-K. Skylaris, P.D. Haynes, A.A. Mostofi, M.C. Payne, Introducing ONETEP: linear-scaling density functional simulations on parallel computers. J. Chem. Phys. **122**, 084119 (2005)
62. P.D. Haynes, C.-K. Skylaris, A.A. Mostofi, M.C. Payne, Elimination of basis set superposition error in linear-scaling density-functional calculations with local orbitals optimised in situ. Chem. Phys. Lett. **422**, 345 (2006)
63. N.W. Ashcroft, N.D. Mermin, *Solid State Physics* (Harcourt Brace College Publishers, Fort Worth, 1976)
64. C. Kittel, *Introduction to Solid State Physics* (Wiley, New York, 2005)
65. L.P. Bouckaert, R. Smoluchowski, E. Wigner, Theory of Brillouin zones and symmetry properties of wave functions in crystals. Phys. Rev. **50**(1), 58 (1936)
66. D.J. Chadi, M.L. Cohen, Special points in the Brillouin zone. Phys. Rev. B **8**(12), 5747 (1973)
67. H.J. Monkhorst, J.D. Pack, Special points for Brillouin-zone integrations. Phys. Rev. B **13**(12), 5188 (1976)
68. G. Makov, M.C. Payne, Periodic boundary conditions in ab initio calculations. Phys. Rev. B **51**(7), 4014 (1995)
69. E. Hernández, M.J. Gillan, Self-consistent first-principles technique with linear scaling. Phys. Rev. B **51**(15), 10157 (1995)
70. A.A. Mostofi, P.D. Haynes, C.-K. Skylaris, M.C. Payne, Preconditioned interative minimisation for linear-scaling electronic structure calculations. J. Chem. Phys. **119**, 8842 (2003)
71. D. Baye, P.-H. Heenen, Generalised meshes for quantum mechanical problems. J. Phys. A Math. Gen. **19**, 2041 (1986)
72. S. Goedecker, Linear scaling electronic structure methods. Rev. Mod. Phys. **71**(4), 1085 (1999)
73. G. Galli, Linear scaling methods for electronic structure calculations and quantum molecular dynamics simulations. Curr. Opin. Solid State Mater. Sci. **1**(6), 864 (1996)
74. P.D. Haynes, C.-K. Skylaris, A.A. Mostofi, M.C. Payne, ONETEP: linear-scaling density-functional theory with local orbitals and plane waves. Phys. Stat. Solidi (b) **243**(11), 2489 (2006)
75. J.M. Soler, E. Artacho, J.D. Gale, A. Garcia, J. Junquera, P. Ordejon, D. Sanchez-Portal, The SIESTA method for ab initio order-N materials simulation. J. Phys. Condens. Matter **14**, 2745 (2002)
76. D.R. Bowler, T. Miyazaki, M.J. Gillan, Recent progress in linear scaling ab initio electronic structure techniques. J. Phys. Condens. Matter **14**(11), 2781 (2002)
77. M.J. Han, T. Ozaki, J. Yu, O(N) LDA+U electronic structure calculation method based on the nonorthogonal pseudoatomic orbital basis. Phys. Rev. B **73**(4), 045110 (2006)

78. F. Mauri, G. Galli, Electronic-structure calculations and molecular-dynamics simulations with linear system-size scaling. Phys. Rev. B **50**(7), 4316 (1994)
79. W. Kohn, Density functional and density matrix method scaling linearly with the number of atoms. Phys. Rev. Lett. **76**(17), 3168 (1996)
80. R. McWeeny, Some recent advances in density matrix theory. Rev. Mod. Phys. **32**(2), 335 (1960)
81. P.D. Haynes, C.-K. Skylaris, A.A. Mostofi, M.C. Payne, Density kernel optimization in the ONETEP code. J. Phys. Condens. Matter **20**(29), 294207 (2008)
82. X.-P. Li, R.W. Nunes, D. Vanderbilt, Density-matrix electronic-structure method with linear system-size scaling. Phys. Rev. B **47**(16), 10891 (1993)
83. R.W. Nunes, D. Vanderbilt, Generalization of the density-matrix method to a nonorthogonal basis. Phys. Rev. B **50**(23), 17611 (1994)
84. M.S. Daw, Model for energetics of solids based on the density matrix. Phys. Rev. B **47**(16), 10895 (1993)
85. J.F. Janak, Proof that $\frac{\partial e}{\partial n_i} = \epsilon_i$ in density-functional theory. Phys. Rev. B **18**(12), 7165 (1978)
86. G.H. Wannier, The structure of electronic excitation levels in insulating crystals. Phys. Rev. **52**(3), 191 (1937)
87. J. des Cloizeaux, Energy bands and projection operators in a crystal: analytic and asymptotic properties. Phys. Rev. **135**(3A), A685 (1964)
88. L. He, D. Vanderbilt, Exponential decay properties of Wannier functions and related quantities. Phys. Rev. Lett. **86**, 5341 (2001)
89. C. Brouder, G. Panati, M. Calandra, C. Mourougane, N. Marzari, Exponential localization of Wannier functions in insulators. Physical Review Letters **98**(4), 046402 (2007)
90. G. Galli, M. Parrinello, Large scale electronic structure calculations. Phys. Rev. Lett. **69**(24), 3547 (1992)
91. N. Hine, P. Haynes, A. Mostofi, C.-K. Skylaris, M. Payne, Linear-scaling density-functional theory with tens of thousands of atoms: expanding the scope and scale of calculations with onetep. Comput. Phys. Commun. **180**(7), 1041 (2009)
92. A. Einstein, Die grundlage der allgemeinen relativitätstheorie. Annalen der Physik **354**, 769 (1916)
93. N. Marzari, D. Vanderbilt, M.C. Payne, Ensemble density-functional theory for ab initio molecular dynamics of metals and finite-temperature insulators. Phys. Rev. Lett. **79**(7), 1337 (1997)
94. F. Mauri, G. Galli, R. Car, Orbital formulation for electronic-structure calculations with linear system-size scaling. Phys. Rev. B **47**(15), 9973 (1993)
95. J. Kim, F. Mauri, G. Galli, Total-energy global optimizations using nonorthogonal localized orbitals. Phys. Rev. B **52**(3), 1640 (1995)
96. P. Ordejón, D.A. Drabold, R.M. Martin, M.P. Grumbach, Linear system-size scaling methods for electronic-structure calculations. Phys. Rev. B **51**(3), 1456 (1995)
97. C.-K. Skylaris, P.D. Haynes, A.A. Mostofi, M.C. Payne, Recent progress in linear-scaling density functional calculations with plane waves and pseudopotentials: the ONETEP code. J. Phys. Condens. Matter **20**, 064209 (2008)

# Chapter 2
# Linear-Scaling DFT + U for Large Strongly-Correlated Systems

Electronic correlation effects, perhaps even more so than large system sizes, have long captivated electronic structure theorists. In this chapter, we seek to tackle both challenges simultaneously, detailing and demonstrating a linear-scaling implementation of an efficacious ab initio method for strongly-correlated materials.

Specifically, we begin by describing the physics of strongly-correlated systems and we discuss the difficulties experienced, and their origins, when exchange-correlation (XC) functionals of the local density approximation type are applied to such materials.

We describe the popular Density Functional Theory + Hubbard model (DFT + $U$) method for overcoming these difficulties, briefly discussing its historical development and motivating it as a corrective idempotency penalty functional of a type frequently employed in linear-scaling DFT methods.

We detail an implementation of DFT + $U$ for which the computational effort for calculation of the ground state energy and forces scales linearly with system size. Expressions for optimising the density and ionic positions are derived in full and in a manner which is applicable to any ab initio approach which employs a set of spatially localised, possibly nonorthogonal, functions to represent the single-particle density matrix. We assume no specific form for the projectors used to define the correlated subspaces in DFT + $U$ and include the necessary adaptations to allow for their nonorthogonality.

## 2.1 Strongly-Correlated Systems

The routine ab initio study of strongly correlated systems, that is those for which the accurate description of the physics is beyond the capacity of band-structure methods such as the unrestricted Hartree–Fock approximation [1], or, somewhat less strictly-speaking, Kohn–Sham DFT [2, 3] within local or semi-local approximations to the XC functional, remains a challenge for electronic structure calculations.

The physics of localised electrons bound to first-row transition metal or lanthanoid ions in such systems is important for understanding and harnessing the behaviour

D. D. O'Regan, *Optimised Projections for the Ab Initio Simulation of Large and Strongly Correlated Systems*, Springer Theses, DOI: 10.1007/978-3-642-23238-1_2, 

of complex systems such as molecular magnets [4], inorganic catalysts [5] and the organometallic molecules that facilitate some of the most critical chemical reactions in biochemistry [6]. Indeed, it is often such physics which is central to the interesting functionality of such materials.

Despite its success at predicting ground-state properties of materials, Kohn–Sham DFT [2, 3] fails to describe the physics of such strongly correlated systems when local or semi-local XC functionals are used, often predicting results that are not only quantitatively but qualitatively inconsistent with experiment.

One example of such a failure is the case of Mott–Hubbard insulating solids [7], characterised by narrow bands of $3d$ or $4f$ orbital orbital character adjacent to the Fermi level; the LSDA [8] may badly underestimate local magnetic moments and may even predict a non-zero density of states at the Fermi level [9, 10].

In order to understand the origin of this deficiency, not least because it serves to motivate the DFT + $U$ method, let us consider the renowned Hubbard model [11–13] for strongly correlated fermionic systems. The Hubbard Hamiltonian is usually written in terms of Fermionic creation, $c_m^{(\sigma)\dagger}$, and annihilation, $c_m^{(\sigma)}$, operators, where it is defined as

$$\hat{H}^{(\sigma)} = \sum_{mm'} t_{mm'} c_m^{(\sigma)\dagger} c_{m'}^{(\sigma)} + \frac{1}{2} \sum_{mm'm''m'''} U_{mm'm''m'''} c_m^{(\sigma)\dagger} c_{m'}^{(-\sigma)\dagger} c_{m'''}^{(-\sigma)} c_{m''}^{(\sigma)}, \tag{2.1}$$

the indices $\{m\}$ labelling sites in which the electrons, with spin index $\sigma$, may reside.

In a continuum model for a real strongly-correlated system, it is useful to use a spatially localised set of single-particle basis orbitals $\{\varphi_m\}$, which we assume here to be spin-independent, with which the creation and annihilation operators, respectively, are spatially resolved via replacement by field operators as per

$$\hat{c}^{(\sigma)}(\mathbf{r}) = \sum_m \varphi_m(\mathbf{r})\, c_m^{(\sigma)}, \quad \text{and} \quad \hat{c}^{(\sigma)\dagger}(\mathbf{r}) = \sum_m \varphi_m^*(\mathbf{r})\, c_m^{(\sigma)\dagger}. \tag{2.2}$$

These basis orbitals may, for example, take the form of Wannier functions constructed from a linear-combination of Bloch states, as described in Chap. 1. Here we assume that the basis is orthonormal, but the generalisation to the nonorthogonal case is available [14].

In the Hubbard model, the Coulomb repulsion between electrons is introduced by the Hubbard $U$ parameter, that is in its orbitally-decomposed form

$$U_{mm'm''m'''} = \int d\mathbf{r} \int d\mathbf{r}'\, \varphi_m^*(\mathbf{r})\, \varphi_{m'}^*(\mathbf{r}')\, \hat{v}(\mathbf{r}, \mathbf{r}')\, \varphi_{m''}(\mathbf{r})\, \varphi_{m'''}(\mathbf{r}').$$

For a given form of interaction $\hat{v}(\mathbf{r}, \mathbf{r}')$, the Hubbard $U$ introduces an energy penalty for occupying nearby orbitals and thus correlates the behaviour of different electrons. The factor of one-half eliminates double-counting over pairs of electrons.

The tendency for the electrons to delocalise, by minimising their kinetic energy, is governed by the simple hopping term

$$t_{mm'} = \int d\mathbf{r}\, \varphi_m^* (\mathbf{r}) \left[ \frac{1}{2} \nabla^2 + \hat{V}_{ext} (\mathbf{r}) \right] \varphi_{m'} (\mathbf{r}), \tag{2.3}$$

which may also include any externally imposed potential.

We will briefly discuss a very simple approximation to the Hubbard model, applied to a simple geometry. Ignoring all but density-density interactions, i.e., $m = m''$, $m' = m'''$; interactions between electrons on one site, i.e., $m = m'$; nearest-neighbour single-particle same-spin hopping terms only, where $m' = m \pm 1$ in one dimension; and identical interaction strengths on each site, we simplify the Hamiltonian to

$$\hat{H}^{(\sigma)} = -t \sum_{\langle mm' \rangle} c_m^{(\sigma)\dagger} c_{m'}^{(\sigma)} + \frac{1}{2} U \sum_m \hat{n}_m^{(\sigma)} \hat{n}_m^{(-\sigma)}. \tag{2.4}$$

Here $\hat{n}_m^{(\sigma)} = c_m^{(\sigma)\dagger} c_m^{(\sigma)}$ measures the occupation of site $m$ with an electron of spin $\sigma$ and the Pauli principle excludes double-occupancy by electrons with identical spins.

Let us apply this model to a periodic chain of sites, for example a one-dimensional chain of $s$-orbitals or Hydrogen atoms. We may consider two limits. In the limit of $U \ll t$, where correlation effects are weak or the atoms lie close together, the reduction of the kinetic energy is the dominant factor and the low-energy eigenstates are made up of delocalised linear combinations of the basis orbitals. There is no strong distinction between the energy terms acting on occupied and unoccupied levels and there is a continuum of states crossing the Fermi level. At the opposite limit, where the Hubbard $U$ or inter-atomic spacing are large, so that $U \gg t$, the minimisation of orbital double occupancy is paramount, the eigenstates become spatially localised on their basis orbitals. In this case, at half-filling, an energy gap of approximately $U \approx I - A$, where, respectively, $I$ and $A$ are the ionisation potential and binding affinity of hydrogen, opens between the occupied and unoccupied levels, in the same way as in a Mott–Hubbard insulator.

Next, let us see how Kohn–Sham DFT may fit into such a framework. The mapping of the interacting many-electron system onto an equivalent system of noninteracting fermions, which are subject an effective single-particle potential, is central to periodic band-structure methods such as Kohn–Sham DFT. In the language of the Hubbard model, we would write the hopping matrix elements of the Kohn–Sham Hamiltonian as

$$t_{mm'}^{DFT} = \int d\mathbf{r}\, \psi_m^* (\mathbf{r}) \left[ \frac{1}{2} \nabla^2 + \hat{V}_{ext} (\mathbf{r}) + \hat{V}_{Hxc} [n] (\mathbf{r}) \right] \psi_{m'} (\mathbf{r}), \tag{2.5}$$

where, in fact, the hopping term makes up the Hamiltonian entirely.

The use of mean-field approximations for the effective potential, $\hat{V}_{Hxc} [n] (\mathbf{r})$, such as the LSDA [8], is appropriate and highly successful in systems where the magnitude of the electron's kinetic energy is large compared with that of the Coulomb interaction between them, so that $U \ll t$ in which case the neglect of explicit Coulomb correlations is justified. In such systems, usually comprising elements whose $3d$ or $4f$ atomic-like states are either completely empty or filled, the electrons are said to be delocalised, or itinerant in extended systems, and are only scattered weakly by atomic centres.

In strongly correlated systems such as Mott–Hubbard insulators, however, the low-dispersion electrons associated with partial occupation of the aforementioned localised $3d$ or $4f$ atomic orbitals do not fall in the regime of $U \ll t$. The LSDA may thus be found to be severely lacking in accuracy due to its lack of explicit Coulomb correlations.

It is clear that within such simple mean-field band-theories, returning to our simple example, that the hydrogen chain spuriously remains metallic as we increase the inter-atomic distance, retaining a diminishing, though finite, density of states at the Fermi level.

The origin of this apparent failure has been understood since the work of Perdew et al. [15] and is related to the unphysical curvature of the energy functional with respect to electronic occupation number [16–18] inherent to LSDA-type functionals unless a self-interaction correction is employed [19]. In the following section, we describe the DFT + $U$ method that, depending on how we wish to look upon it, reintroduces explicit Coulomb interactions to the Kohn–Sham Hamiltonian or reintroduces the appropriate derivative discontinuity to the XC functional.

## 2.2 The DFT + $U$ Method

A number of sophisticated methods to correct the description of strong correlation effects within Kohn–Sham DFT have been developed which provide a good compromise between accuracy and computational expense. Many of these methods, notably DFT +Hubbard $U$ (DFT + $U$) [20, 21] and DFT + dynamical mean field theory (DFT + DMFT) [22, 23] for static and dynamical spatially localised Coulomb correlation effects, respectively, share a common history and conceptual motivation which is based on the Hubbard model we have discussed. In such methods, the electronic system is subdivided into a set of spatially localised correlated subspaces and the remainder which acts as a bath for particle exchange. The description of the strong Coulomb interactions, i.e., $U \geq t$, between particles in the correlated subspaces is deemed to be beyond the capacity of the XC functional, so that explicit supplementation using the Hubbard model is required. In the remainder of the system, the kinetic energy is supposed to be large relative to Coulomb interaction, i.e., $U \ll t$, and the XC functional is assumed to perform adequately. In this manner, a Hubbard model interaction may be used to augment the description of the screened Coulomb interactions in the correlated subspaces while retaining the computationally inexpensive mean-field model for the free-electron like remainder of the system.

Figure 2.1 illustrates, as an example, a spatially delocalised single-particle orbital, $\psi_n(\mathbf{r})$, together with a localised Wannier function (technically an NGWF), $\phi_m(\mathbf{r})$,

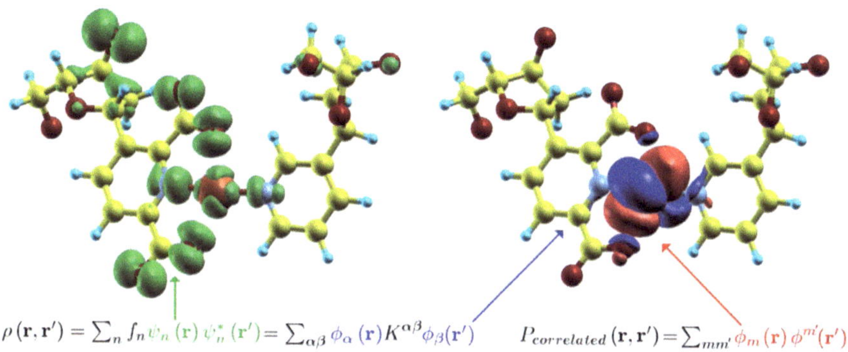


**Fig. 2.1** Detail of a $Cu^{2+}$-mediated DNA base pair [24] in the system on the right of Fig. 1.1, showing the delocalised highest occupied majority-spin molecular orbital (*left*) for comparison with a localised NGWF of localised 3$d$-orbital character (*right*)

used to represent it in an artificial molecular system of technological interest. The description of the Coulomb interactions within the partially filled $Cu^{2+}(3d)$ sub-shell of such systems may benefit from the use of DFT + $U$.

We will henceforth describe the orbitals used to delineate the strongly correlated subspaces as *Hubbard projectors*. These are typically spatially localised on a particular transition-metal or lanthanoid atom and usually, but not necessarily, of the same number per correlated site as the number of orbitals, $2l^{(I)} + 1$, in the most localised hydrogenic valence sub-shell of the atom at that site, e.g., we use five Hubbard projectors for a 3$d$ sub-shell. Localised Wannier functions built from the Kohn–Sham eigenfunctions may offer an efficient set of Hubbard projectors, and we discuss this possibility further in Chap. 3 and Ref. [25].

In DFT + $U$, consistency between the subspaces and the bath is provided by ensuring that the electronic density-matrix for the complete system remains subject to the usual requirements of idempotency, compatibility with the ground-state Hamiltonian and proper normalisation. In DFT + DMFT, on the other hand, self-consistency over the density is not routinely enforced at present, although successful examples of such calculations have been demonstrated [26]. The equivalence of the correlated subspace Green's functions and the projection of the full Green's function onto these subspaces is, however, required.

Generally for these methods, the occupancy matrix of each correlated subspace is the object which provides, for a given set of Hubbard $U$ parameters which may or may not depend on the density and its response, the necessary information on the electronic density-matrix to the Hubbard model which describes intra-subspace interactions. In Chap. 3, published in Ref. [25], we will describe a self-consistent method for delineating the correlated subspaces based on Wannier functions; in Chap. 4, published in Ref. [27], we discuss the definition of subspace occupancy matrices when using nonorthogonal projector functions; and in Chap. 7 we address the computation of the Hubbard $U$ parameters in the nonorthogonal formalism.

Let us restrict ourselves initially, for simplicity, to the case of orthonormal Hubbard projectors labelled $\{m\}$ at correlated site $I$, for which the occupancy matrix of spin $\sigma$ electrons is given by

$$n_{mm'}^{(I)(\sigma)} = Tr\left[c_{m'}^{(I)\dagger} c_{m}^{(I)\dagger} \hat{\rho}^{(\sigma)}\right] = \langle \varphi_{m}^{(I)} | \hat{\rho}^{(\sigma)} | \varphi_{m'}^{(I)} \rangle. \tag{2.6}$$

Considering, as our starting point, the first rotationally-invariant form of DFT + $U$ introduced in the literature, that of Refs. [28, 29] and known as LSDA + $U$. In this method, the correlation, classical Coulomb repulsion and exchange parts of the fermionic Hubbard Hamiltonian, Eq. 2.1, are expressed separately for each spatially localised correlated subspace. The trace of these interaction terms with the appropriate occupancy matrices gives the energy expectation value of the Hubbard Hamiltonian, for each subspace, with the Kohn–Sham density-matrix.

Summation over correlated sites and the electron spin index gives the energy correction due to explicit Coulomb interactions for this density-matrix, which is given by

$$\begin{aligned} E_U = \frac{1}{2} \sum_{I\sigma\{m\}} \Big\{ & U_{mm''m'm'''}^{(I)} n_{mm'}^{(I)(\sigma)} n_{m''m'''}^{(I)(-\sigma)} \\ & + \left( U_{mm''m'm'''}^{(I)} - U_{mm''m'''m'}^{(I)} \right) n_{mm'}^{(I)(\sigma)} n_{m''m'''}^{(I)(\sigma)} \Big\}. \end{aligned} \tag{2.7}$$

Here, the first, second and third terms correspond, respectively, to spin off-diagonal density-density repulsion (correlation), spin-diagonal density-density repulsion and spin-diagonal exchange effects. If unscreened Coulomb interactions are used to build the Hubbard $U$ parameters, this is the Hartree-Fock approximation to the Coulomb energy of the correlated subspaces.

The contribution to the DFT energy functional arising from the correlated subspaces and already included in the conventional exchange correlation term in a mean-field sense must be subtracted in order to approximately remove double-counting of the Coulomb interactions. The DFT + $U$ energy functional is thus generally given by

$$E_{DFT+U} = E_{DFT} + E_U - E_{DC}. \tag{2.8}$$

The double-counting term used in this rotationally-invariant form of LSDA + $U$ model is the simple "atomic limit" approximation detailed in Ref. [30], calculated by presupposing an integer occupancy of the correlated subspaces and thus given by

$$\begin{aligned} E_{DC} = \frac{1}{2} \sum_{I\sigma} \Big\{ & U^{(I)} N^{(I)(\sigma)} \left( \sum_{\sigma'} N^{(I)(\sigma')} - 1 \right) \\ & - J^{(I)} N^{(I)(\sigma)} \left( N^{(I)(\sigma)} - 1 \right) \Big\}. \end{aligned} \tag{2.9}$$

Here, the total occupancy for a given site and spin and subspace-averaged Coulomb repulsion and spin-diagonal exchange parameters are denoted, respectively, by

$$N^{(I)(\sigma)} = \sum_{m} N_{mm}^{(I)(\sigma)}, \tag{2.10}$$

$$U^{(I)} = \frac{1}{\left(2l^{(I)}+1\right)^2} \sum_{m,m'} U_{mm'mm'}^{(I)} \quad \text{and} \tag{2.11}$$

$$J^{(I)} = \frac{1}{\left(2l^{(I)}\right)\left(2l^{(I)}+1\right)} \sum_{m,m'\neq m} U_{mm'm'm}^{(I)}. \tag{2.12}$$

Alternative forms of the double-counting correction, such as those described in Refs. [20, 31], are also available although we do not discuss them further.

Following Ref. [16], which offers a simplified rotationally-invariant DFT + $U$ functional which itself is based on previous proposals in Ref. [10], we next neglect the corrections associated with exchange effects and with interactions between electrons of different spin. Some manipulation allows us to re-write the DFT + $U$ correction, in this approximation, as

$$\begin{aligned} E_{DFT+U} &= \frac{1}{2} \sum_{I\sigma} \Bigg\{ \sum_{mm'} U_{mm'mm'}^{(I)} \left( n_{mm}^{(I)(\sigma)} n_{m'm'}^{(I)(\sigma)} - n_{mm'}^{(I)(\sigma)} n_{m'm}^{(I)(\sigma)} \right) \\ &\quad - U^{(I)} N^{(I)(\sigma)} \left( N^{(I)(\sigma)} - 1 \right) \Bigg\} \end{aligned} \tag{2.13}$$

$$\begin{aligned} &= \frac{1}{2} \sum_{I\sigma} \sum_{mm'} \Big\{ \left( U_{mm'mm'}^{(I)} - U^{(I)} \right) n_{mm}^{(I)(\sigma)} n_{m'm'}^{(I)(\sigma)} \\ &\quad + U^{(I)} n_{mm'}^{(I)(\sigma)} \delta_{mm'} - U_{mm'mm'}^{(I)} n_{mm'}^{(I)(\sigma)} n_{m'm}^{(I)(\sigma)} \Big\}. \end{aligned} \tag{2.14}$$

Finally, we may simplify this further by approximating the orbitally-decomposed $U_{mm'mm'}^{(I)}$ by its average scalar $U^{(I)}$, in which case the first term in Eq. 2.14 vanishes to give the widely-used simplified DFT + $U$ functional

$$\begin{aligned} E_{DFT+U} &= \frac{1}{2} \sum_{I\sigma} \sum_{mm'} U^{(I)} \left( n_{mm'}^{(I)(\sigma)} \delta_{mm'} - n_{mm'}^{(I)(\sigma)} n_{m'm}^{(I)(\sigma)} \right) \\ &= \frac{1}{2} \sum_{I\sigma} \sum_{m} U^{(I)} \left( n_{mm}^{(I)(\sigma)} - \sum_{m'} n_{mm'}^{(I)(\sigma)} n_{m'm}^{(I)(\sigma)} \right). \end{aligned} \tag{2.15}$$

The interaction averaging approximation, which is expected to be most valid for a spherically symmetric correlated subspace immersed in an isotropic environment,

$$U^{(I)}_{mm'mm'} \approx U^{(I)}, \quad \forall\, m, m' \in \left\{1, \cdots, 2l^{(1)} + 1\right\}, \tag{2.16}$$

might suggest that we may alternatively replace the scalar approximation $U^{(I)}$ by its orbital dependent counterpart in Eq. 2.14. This also gives a rather simple DFT + $U$ functional,

$$E_{DFT+U} = \frac{1}{2} \sum_{I\sigma} \sum_{mm'} U^{(I)}_{mm'mm'} \left( n^{(I)(\sigma)}_{mm} \delta_{mm'} - n^{(I)(\sigma)}_{mm'} n^{(I)(\sigma)}_{m'm} \right),$$

however it is not of practical utility since it is not invariant under unitary rotations among the Hubbard projectors. We return to the question of constructing tensorially valid DFT + $U$ corrections with orbital-dependent parameters in Chap. 7.

The principal effect of the DFT + $U$ correction Eq. 2.15 is to approximately emulate the exact XC functional by introducing a derivative discontinuity in the total-energy with respect to the occupancy matrix of the correlated subspaces at integer values. The DFT + $U$ correction to the Kohn–Sham potential, given by

$$\hat{V}_{DFT+U} = \sum_{I\sigma} \sum_{mm'} U^{(I)} |\varphi^{(I)}_m\rangle \left( \frac{1}{2}\delta_{mm'} - n^{(I)(\sigma)}_{mm'} \right) \langle \varphi^{(I)}_{m'}|, \tag{2.17}$$

acts to restore the correct occupancy dependence of the potential, and is attractive or repulsive for occupancy-matrix elements greater or less than one-half, respectively. The result, to a first approximation, is the penalisation of non-integer correlated subspace occupancies and, consequently, an opening of an energy gap of order $U$ between the occupied and unoccupied Kohn–Sham states which have a large overlap with the Hubbard projectors, thus facilitating the simulation of strongly-correlated systems such as Mott–Hubbard insulators within Kohn–Sham DFT.

The DFT + $U$ correction to the energy functional allows a simple interpretation, or perhaps motivation, as an idempotency penalty-functional [32] of the type often used to maintain the idempotency of the Kohn–Sham density-matrix in linear-scaling DFT. Assuming that the particles occupying the correlated subspaces interact strongly with each other, compared to their interaction with the bath, each subspace effectively acts as an individual open quantum system. As such, we could separately impose the density-matrix idempotency condition, i.e., Fock antisymmetry, of the projected density-matrix for each subspace.

However, the idempotency of the density-matrix for the complete Kohn–Sham system is a condition which must be exactly satisfied, at the ground-state, and the idempotency of each subspace density-matrix is, in general, a competing condition. Thus, the subspace idempotency may be only partially enforced up to an idempotency functional of the form

$$\sum_{I\sigma} \mathbf{Tr} \left[ \lambda^{(I)(\sigma)} \left( \hat{n}^{(I)(\sigma)} - \hat{n}^{(I)(\sigma)2} \right) \right]; \quad \lambda^{(I)(\sigma)} = \frac{U^{(I)(\sigma)}}{2}, \tag{2.18}$$

which penalises the degradation of fermionic behaviour in each correlated subspace. Since the strength of the effective Coulomb interactions is closely related to the extent to which the derivative discontinuity is lacking in the XC functional, and the latter is responsible for the spurious partial-occupancy of localised states (or correlated subspace idempotency deviation) by that functional, the Hubbard $U$ parameter may be heuristically identified with the pre-factor of the idempotency penalty-functional.

## 2.3 Framework for Linear-Scaling DFT + $U$

We now proceed to describe the implementation of the DFT + $U$ functional defined by Eq. 2.15 in a contemporary approach to linear-scaling DFT. Firstly, in this section, we establish a notational framework for expressing the Hubbard projectors, which are permitted to be nonorthogonal for each correlated subspace, in terms of a localised nonorthogonal set of support functions of the type typically used to represent the Kohn–Sham density matrix in many linear-scaling approaches.

Due to the spatial localisation of both the Hubbard projectors and the support functions, matrix sparsity patterns naturally play an important role in the construction of our linear-scaling DFT + $U$ method. In fact, as we will show, matrix sparsity patterns allow us to carry out DFT + $U$ calculations involving a large number of correlated subspaces in a very efficient manner.

We will make due comment on matrix sparsity issues, when appropriate, as we describe the elements of our linear-scaling implementation of DFT + $U$. Linear scaling with respect to the number of correlated subspaces in the system may be achieved for some elements of the DFT + $U$ module, and, in less favourable cases, with respect to the total number of atoms in the simulation cell.

We begin by expressing the Hubbard projectors $\{\varphi_m^{(I)}\}$ spanning each correlated subspace $I$ in terms of the nonorthogonal basis functions $\{\phi_\alpha\}$ via the linear transformation

$$\varphi_m^{(I)}(\mathbf{r}) = \phi_\alpha(\mathbf{r})\, S^{\alpha\beta} V_{\beta m}^{(I)}. \tag{2.19}$$

Here, we must assume that the Hubbard projectors are fully expandable in the frame of their surrounding NGWFs, though this does not introduce any limitation, in practice, since the explicit expansion of the projectors in the *psinc* function basis is used in the update of the NGWFs themselves. We also assume, for notational simplicity, that identical Hubbard projectors are used for each spin channel, although the generalisation to spin-dependent projectors is straightforward.

Here $S^{\alpha\beta} = \langle \phi^\alpha | \phi^\beta \rangle$ is the contravariant metric on the NGWFs, as usual, and it follows that the transformation matrix between covariant basis functions and projectors is given by

$$V_{\beta m}^{(I)} = \langle \phi_\beta | \varphi_m^{(I)} \rangle, \tag{2.20}$$

which may be a very sparse matrix for a low density of correlated subspaces.

It will be convenient to establish the adjoint (once covariant for the support function index and once for the Hubbard projector index) transformation matrix as

$$W^{(I)}_{m\alpha} = V^{(I)\dagger}_{\alpha m} = \langle \varphi^{(I)}_m | \phi_\alpha \rangle. \tag{2.21}$$

The metric on each correlated subspace is discussed in detail in Chap. 4 and Ref. [27]; most of the expressions in this chapter extend readily to the delocalised Hubbard projector duals discussed there. Here, for brevity, we will restrict ourselves to the case of the localised Hubbard projector duals defined by

$$|\varphi^{(I)m}\rangle = |\varphi^{(I)}_{m'}\rangle O^{(I)m'm}; \quad O^{(I)}_{mm'} = \langle \varphi^{(I)}_m | \varphi^{(I)}_{m'} \rangle, \tag{2.22}$$

where an individual metric tensor $O^{(I)}_{\bullet\bullet}$ is used for each correlated subspace, as we propose in Chap. 4 and Ref. [27], in order to maintain the tensorial invariance of the total energy.

The generalised occupation matrix for each correlated subspace, a mixed tensor with respect to the projector indices, is expressed in the support function representation as

$$\begin{aligned} n^{(I)(\sigma)m'}_{\ \ m} &= \langle \varphi^{(I)}_m | \hat{\rho}^{(\sigma)} | \varphi^{(I)m'} \rangle \\ &= \langle \varphi^{(I)}_m | \phi_\alpha \rangle K^{(\sigma)\alpha\beta} \langle \phi_\beta | \varphi^{(I)}_{m''} \rangle O^{(I)m''m'} \\ &= W^{(I)}_{m\alpha} K^{(\sigma)\alpha\beta} V^{(I)}_{\beta m''} O^{(I)m''m'} \\ &= \left( W^{(I)} K^{(\sigma)} V^{(I)} O^{(I)} \right)_m^{\ \ m'}, \end{aligned} \tag{2.23}$$

where

$$K^{(\sigma)\alpha\beta} = \sum_{i\mathbf{k}} M^{(\sigma)\alpha}{}_{i\mathbf{k}} f^{(\sigma)}_{i\mathbf{k}} (M^{(\sigma)\dagger})_{i\mathbf{k}}{}^{\beta} \tag{2.24}$$

is the density kernel relating support functions (assumed to be spin-independent) to Kohn–Sham orbitals via a linear transformation matrix $M^{(\sigma)\alpha}{}_{i\mathbf{k}} = \langle \phi^\alpha | \psi^{(\sigma)}_{i\mathbf{k}} \rangle$.

The DFT + $U$ correction to the total energy, in the case of a generalised spin-dependent two-index interaction tensor $U^{(I)(\sigma)m'}_{\ \ m}$ (approximations for the interaction tensor of various rank are discussed in Chap. 7), is generally computed using the matrix trace

$$\begin{aligned} E_{DFT+U} &= \sum_{I,\sigma} \frac{1}{2} U^{(I)(\sigma)m'}_{\ \ m} \left[ n^{(I)(\sigma)m''}_{\ \ m'} \left( \delta_{m''}^{\ m} - n^{(I)(\sigma)m}_{\ \ m''} \right) \right] \\ &= \sum_{I,\sigma} \frac{1}{2} Tr \left[ \begin{array}{l} U^{(I)} W^{(I)} K^{(\sigma)} V^{(I)} O^{(I)} \\ \times \left( 1 - W^{(I)} K^{(\sigma)} V^{(I)} O^{(I)} \right) \end{array} \right], \end{aligned} \tag{2.25}$$

where here, with further examples to follow, we use multiple continued lines within braces to describe lengthy scalar expressions.

Of course, it is undesirable, both from the point of view of implementation and computational efficiency, to explicitly use separate $V^{(I)}$, $W^{(I)}$ and $O^{(I)}$ matrices for each site. This requires individual matrix products to be carried out for each site before the sum over sites is computed. If care is taken, however, we may safely embed all of these small matrices into large, very sparse, $V$, $W$ and $O$ matrices for the entire system. These large matrices then fit seamlessly into the hierarchical, parallelised, sparse algebra routines found in a modern linear-scaling DFT code such as ONETEP.

Let us analyse this strategy in greater detail, taking as an example the computation of the occupancy matrix

$$n^{(I)(\sigma)m'}_{m} = W^{(I)}_{m\alpha} K^{(\sigma)\alpha\beta} V^{(I)}_{\beta m''} O^{(I)m''m'}. \tag{2.26}$$

Working from right to left and temporarily placing a site index before each projector index to clarify its meaning, first consider the product

$$(VO)^{(I)m'}_{\beta} = \sum_{J} V_{\beta(J)m''} O^{(\delta_{JI} J)m''(I)m'} = V_{\beta(I)m''} O^{(I)m''(I)m'}, \tag{2.27}$$

which retains the same sparsity pattern as $V$ due to the block-sparsity of the $O$ matrix (the size of each block is the number of projectors spanning the subspace on the site in question).

Next, taking the product from the left with the density kernel,

$$(KVO)^{(\sigma)\alpha(I)m'} = K^{(\sigma)\alpha\beta} (VO)^{(I)m'}_{\beta}, \tag{2.28}$$

we see that this too has the same sparsity as $V$ when no density kernel truncation is applied, in which case the indices $\alpha$ and $\beta$ run over all NGWFs. When kernel truncation is enforced, however, the number of values which $\alpha$ can take is reduced and the effort needed for the sum over $\beta$ is diminished.

Only on the final step, where we compute

$$n^{(\sigma)(I)m'}_{(J)m} = W_{(J)m\alpha} (KVO)^{(\sigma)\alpha(I)m'}, \tag{2.29}$$

do we accumulate extraneous information on the off-site non-locality of the density matrix. Were we to compute this matrix in full and then consider its square, for example, we would find that

$$\sum_{K} n^{(\sigma)(K)m''}_{(I)m} n^{(\sigma)(I)m'}_{(K)m''} \neq n^{(\sigma)(I)m'}_{m} n^{(\sigma)(I)m'}_{m}. \tag{2.30}$$

The former is what is generated in the matrix product, while the latter is what we require. This problem is resolved by explicitly truncating the required occupancy matrix

$$n^{(\sigma)(I)m'}_{(I)m} = W_{(I)m\alpha} (KVO)^{(\sigma)\alpha(I)m'} \tag{2.31}$$

to the same sparsity pattern as $O$ (which is also the same as that of $U$), thus eliminating any unwanted off-site occupancies. In practice, the unnecessary elements are never actually computed, and no wasted effort is incurred, since the sparse algebra system takes into consideration the sparsity pattern of the product matrix.

## 2.4 Variations with Respect to the Density Kernel

In the ONETEP code, and indeed most current linear-scaling DFT methods, the LNVD [33–35] technique, described in Chap. 1, is used to minimise the energy with respect to the density-matrix, bringing it closer to commutativity with the Kohn–Sham Hamiltonian while simultaneously driving it towards idempotency. This takes place in the inner energy minimisation loop in ONETEP, where the NGWFs are kept fixed and a non-linear conjugate gradients minimisation of the energy with respect to the matrix elements of the LNVD auxiliary density kernel, $L^{(\sigma)\alpha\beta}$, is carried out.

The Kohn–Sham density kernel is related to the auxiliary density kernel via one iteration of the McWeeny purification transform, i.e.,

$$K^{(\sigma)\alpha\beta} = (3LSL - 2LSLSL)^{(\sigma)\alpha\beta} \,. \tag{2.32}$$

In our treatment of DFT + $U$, we go a step further and provide the more general expressions needed for the HSMP [36] adaptation of the LNVD method, in which a density kernel $\tilde{K}^{(\sigma)}$ is expressed as a purified and normalised auxiliary density kernel, explicitly

$$\tilde{K}^{(\sigma)\alpha\beta} = \frac{K^{(\sigma)\alpha\beta} N^{(\sigma)}}{K^{(\sigma)\gamma\delta} S_{\delta\gamma}} = \frac{(3LSL - 2LSLSL)^{(\sigma)\alpha\beta} \, N^{(\sigma)}}{(3LSL - 2LSLSL)^{(\sigma)\gamma\delta} \, S_{\delta\gamma}}, \tag{2.33}$$

where $N^{(\sigma)}$ is the correct occupancy of spin channel $\sigma$. The kernel renormalisation introduces terms in the gradient akin to a chemical potential, which project out any first-order changes to the electron number, driving the density kernel $\tilde{K}^{(\sigma)}$ towards both normalisation and idempotency as the energy is minimised.

To locate the doubly-covariant derivative of the DFT + $U$ energy term with respect to the auxiliary density kernel, stressing that it is computed strictly using the purified and renormalised density kernel, we make use of the chain-rule for matrix derivatives to write (suppressing the spin index for concision)

$$\frac{\partial E_{DFT+U}}{\partial L^{\alpha\beta}} = \frac{\partial E_{DFT+U}}{\partial K^{\iota\kappa}} \frac{\partial K^{\iota\kappa}}{\partial L^{\alpha\beta}}. \tag{2.34}$$

It may be readily shown that the latter term is given by

$$\begin{aligned} \frac{\partial K^{\iota\kappa}}{\partial L^{\alpha\beta}} &= 3\left(\delta^{\iota}_{\alpha} S_{\beta\gamma} L^{\gamma\kappa} + L^{\iota\gamma} S_{\gamma\alpha} \delta^{\kappa}_{\beta}\right) \\ &- 2\left( \begin{array}{c} \delta^{\iota}_{\alpha} S_{\beta\gamma} L^{\gamma\eta} S_{\eta\xi} L^{\xi\kappa} \\ + L^{\iota\gamma} S_{\gamma\alpha} S_{\beta\eta} L^{\eta\kappa} + L^{\iota\gamma} S_{\gamma\eta} L^{\eta\xi} S_{\xi\alpha} \delta^{\kappa}_{\beta} \end{array} \right). \end{aligned} \tag{2.35}$$

The derivative of the DFT + $U$ energy term with respect to the purified density kernel $K^{\bullet\bullet}$ may be broken into products of derivatives and rearranged as follows

$$
\begin{aligned}
\frac{\partial E_{DFT+U}}{\partial K^{\iota\kappa}} &= \frac{\partial}{\partial \tilde{K}^{\eta\theta}} \left[ E_{DFT+U} \left( \tilde{K}^{\bullet\bullet} \right) \right] \frac{\partial \tilde{K}^{\theta\eta}}{\partial K^{\iota\kappa}} \\
&= H_{\eta\theta}^{DFT+U} \frac{\partial}{\partial K^{\iota\kappa}} \left[ \frac{N}{K^{\alpha\beta} S_{\beta\alpha}} K^{\theta\eta} \right] \\
&= H_{\eta\theta}^{DFT+U} \frac{N}{\left( K^{\alpha\beta} S_{\beta\alpha} \right)^2} \left[ \frac{\partial K^{\theta\eta}}{\partial K^{\iota\kappa}} \left( K^{\gamma\delta} S_{\delta\gamma} \right) - K^{\theta\eta} S_{\iota\kappa} \right] \\
&= H_{\eta\theta}^{DFT+U} \frac{N}{\left( K^{\alpha\beta} S_{\beta\alpha} \right)} \left[ \delta_\theta^\iota \delta_\eta^\kappa - \frac{K^{\theta\eta}}{\left( K^{\gamma\delta} S_{\delta\gamma} \right)} S_{\iota\kappa} \right].
\end{aligned} \tag{2.36}
$$

We may next write the gradient with respect to the density kernel in terms of a preconditioned contribution to the Hamiltonian, denoted by $\tilde{H}_{\bullet\bullet}^{DFT+U}$, identifying the DFT + $U$ correction to the chemical potential needed to preserve the electron number of the system as

$$
\mu^{DFT+U} = \frac{H_{\eta\theta}^{DFT+U} K^{\theta\eta}}{\left( K^{\gamma\delta} S_{\delta\gamma} \right)}, \tag{2.37}
$$

since

$$
\begin{aligned}
\frac{\partial E_{DFT+U}}{\partial K^{\iota\kappa}} &= \frac{N}{\left( K^{\alpha\beta} S_{\beta\alpha} \right)} \left[ H_{\iota\kappa}^{DFT+U} - \frac{H_{\eta\theta}^{DFT+U} K^{\theta\eta}}{\left( K^{\gamma\delta} S_{\delta\gamma} \right)} S_{\iota\kappa} \right] \\
&= \frac{N}{\left( K^{\alpha\beta} S_{\beta\alpha} \right)} \left[ H_{\iota\kappa}^{DFT+U} - \mu^{DFT+U} S_{\iota\kappa} \right] \\
&= \frac{N}{\left( K^{\alpha\beta} S_{\beta\alpha} \right)} \tilde{H}_{\iota\kappa}^{DFT+U}.
\end{aligned} \tag{2.38}
$$

The DFT + $U$ contribution to the Hamiltonian, used in the above and which is computed using the purified and renormalised density kernel, in practice is given by

$$
\begin{aligned}
H_{\iota\kappa}^{DFT+U} &= \frac{\partial E_{DFT+U}}{\partial K^{\iota\kappa}} \\
&= \sum_I \frac{1}{2} U_{m'}^{(I)m} \left[ \frac{\partial n_m^{(I)m'}}{\partial K^{\iota\kappa}} - \frac{\partial n_m^{(I)m''}}{\partial K^{\iota\kappa}} n_{m''}^{(I)m'} - n_m^{(I)m''} \frac{\partial n_{m''}^{(I)m'}}{\partial K^{\iota\kappa}} \right].
\end{aligned} \tag{2.39}
$$

In order to express this in terms of the NGWF representation, we begin by noting that the partial derivative of the occupation matrix, for a given subspace, with respect to an arbitrary density kernel $K^{\bullet\bullet}$, is given by

$$
\begin{aligned}
\frac{\partial n_m^{(I)m'}}{\partial K^{\iota\kappa}} &= \frac{\partial}{\partial K^{\iota\kappa}} \left[ W_{m\gamma}^{(I)} K^{\gamma\delta} V_{\delta m''}^{(I)} O^{(I)m''m'} \right] \\
&= W_{m\gamma}^{(I)} \delta_\iota^\gamma \delta_\kappa^\delta V_{\delta m''}^{(I)} O^{(I)m''m'} \\
&= W_{m\iota}^{(I)} V_{\kappa m''}^{(I)} O^{(I)m''m'}.
\end{aligned} \tag{2.40}
$$

The trace of this over the Hubbard projectors gives the covariant support function representation of the Hubbard projection for subspace $\mathcal{C}^{(I)}$, that is

$$\sum_m \frac{\partial n^{(I)m}_{\ \ m}}{\partial K^{\iota\kappa}} = V^{(I)}_{\kappa m''} O^{(I)m''m} W^{(I)}_{m\iota} \equiv P^{(I)}_{\kappa\iota}. \tag{2.41}$$

It follows that the products of the occupancy matrix and its derivative, always computed in the frame of Hubbard projectors in practice since there they each have the block-diagonal sparsity pattern of $O$, are expressed in the support function representation as

$$\begin{aligned}\frac{\partial n^{(I)m''}_{\ \ m}}{\partial K^{(\sigma)\iota\kappa}} n^{(I)m'}_{\ \ m''} &= \left(W^{(I)}_{m\iota} V^{(I)}_{\kappa m'''} O^{(I)m'''m''}\right)\left(W^{(I)} K V^{(I)} O^{(I)}\right)_{m''}^{\ \ m'} \\ &= W^{(I)}_{m\iota}\left(P^{(I)} K V^{(I)} O^{(I)}\right)_{\kappa}^{\ \ m'}\end{aligned} \tag{2.42}$$

and, taking the complimentary product,

$$\begin{aligned}n^{(I)m''}_{\ \ m} \frac{\partial n^{(I)m'}_{\ \ m''}}{\partial K^{(\sigma)\iota\kappa}} &= \left(W^{(I)} K V^{(I)} O^{(I)}\right)_{m}^{\ \ m''}\left(W^{(I)}_{m''\iota} V^{(I)}_{\kappa m'''} O^{(I)m'''m'}\right) \\ &= \left(W^{(I)} K P^{(I)}\right)_{m\iota}\left(V^{(I)} O^{(I)}\right)_{\kappa}^{\ \ m'}.\end{aligned} \tag{2.43}$$

We are free to evaluate the Hubbard $U$ interaction operator as a mixed tensor in the NGWF representation, so that

$$U^{(I)\gamma}_{\alpha} = V^{(I)}_{\alpha m} O^{(I)mm'} U^{(I)m''}_{\ \ m'} W^{(I)}_{m''\beta} S^{\beta\gamma}. \tag{2.44}$$

As a result, noting that the Hubbard projection operator and the Hubbard interaction operator commute but that the density-matrix and the Hubbard interaction operator may not, the DFT + $U$ term in the covariant Hamiltonian, denoted $H^{(\sigma)DFT+U}_{\bullet\bullet}$, is expressed in the support function representation by

$$H^{(\sigma)DFT+U}_{\alpha\beta} = \sum_I \frac{1}{2}\begin{pmatrix} U^{(I)} P^{(I)} \\ -P^{(I)} K^{(\sigma)} U^{(I)} P^{(I)} \\ -U^{(I)} P^{(I)} K^{(\sigma)} P^{(I)} \end{pmatrix}_{\alpha\beta}. \tag{2.45}$$

The DFT + $U$ term in the total-energy, on the other hand, is succinctly expressed as

$$E^{(\sigma)}_{DFT+U} = \sum_I \frac{1}{2} U^{(I)\gamma}_{\alpha}\left(P^{(I)} K^{(\sigma)} - P^{(I)} K^{(\sigma)} P^{(I)} K^{(\sigma)}\right)_{\gamma}^{\ \ \alpha}. \tag{2.46}$$

The associated DFT + $U$ independent-particle, "band-structure", energy correction, $E^{IP}_{DFT+U} = H^{DFT+U}_{\iota\kappa} \tilde{K}^{\kappa\iota}$, does not equal the energy term $E_{DFT+U}$ and so the energy correction entering into the computation of $\mu^{DFT+U}$ is not the same as the correction to the total-energy. This explicitly demonstrates that DFT + $U$ is a correction beyond the independent-particle approximation.

The required DFT + $U$ contribution to the covariant gradient is then provided by the product of the preconditioned term in the Hamiltonian and the derivative of the density kernel with respect to its auxiliary counterpart. We find that this is given by (again suppressing the spin index)

$$\begin{aligned}
\frac{\partial E_{DFT+U}}{\partial L^{\alpha\beta}} &= \frac{N}{\left(K^{\gamma\delta} S_{\delta\gamma}\right)} \tilde{H}^{DFT+U}_{\kappa\iota} \frac{\partial K^{\iota\kappa}}{\partial L^{\alpha\beta}} \\
&= \frac{N}{(3LSL - 2LSLSL)^{\gamma\delta} S_{\delta\gamma}} \\
&\times \begin{pmatrix} 3\left(\tilde{H}LS + SL\tilde{H}\right)_{\alpha\beta} \\ -2\left(\tilde{H}LSLS + SL\tilde{H}LS + SLSL\tilde{H}\right)_{\alpha\beta} \end{pmatrix}.
\end{aligned} \tag{2.47}$$

Here, $\tilde{H}_{\bullet\bullet}$ is shorthand for the preconditioned DFT + $U$ term in the Hamiltonian and is given by

$$\tilde{H}^{DFT+U}_{\bullet\bullet} = H^{DFT+U}_{\bullet\bullet} - \mu^{DFT+U} S_{\bullet\bullet}. \tag{2.48}$$

Explicitly, we may now write the DFT + $U$ correction to the Hamiltonian as

$$\tilde{H}^{(\sigma)DFT+U}_{\alpha\beta} = \sum_I \frac{1}{2} U^{(I)\zeta}_{\eta} \begin{bmatrix} P_{\zeta\beta}\left(\delta - PK\right)_{\alpha}{}^{\eta} - (PKP)_{\zeta\beta}\,\delta^{\eta}_{\alpha} \\ -\left(PK - 2PKPK\right)_{\zeta}{}^{\eta} S_{\alpha\beta} \\ \times \left(K^{\gamma\delta} S_{\delta\gamma}\right)^{-1} \end{bmatrix}^{(I)(\sigma)}.$$

For a refinement of the auxiliary density kernel $L^{\alpha\beta}$, any update must also be a contravariantly transforming tensor, as noted in Refs. [14, 37]. In order to provide such a search direction, it is necessary that we pre- and post-multiply the above covariant gradient with the contravariant metric tensor on the NGWFs, that is the inverse overlap matrix of the NGWFs at the point at which the gradient itself is computed, to give

$$G^{(\sigma)\alpha\beta}_{DFT+U} = \left(S^{-1}\right)^{\alpha\gamma} \frac{\partial E_{DFT+U}}{\partial L^{(\sigma)\gamma\delta}} \left(S^{-1}\right)^{\delta\beta} \tag{2.49}$$

Carrying this out, we obtain the DFT + $U$ contribution to the contravariant density kernel gradient,

$$\begin{aligned}
G^{(\sigma)\alpha\beta}_{DFT+U} &= \frac{N}{(3LSL - 2LSLSL)^{\gamma\delta} S_{\delta\gamma}} \\
&\times \begin{pmatrix} 3\left(S^{-1}\tilde{H}L + L\tilde{H}S^{-1}\right) \\ -2\left(S^{-1}\tilde{H}LSL + L\tilde{H}L + LSL\tilde{H}S^{-1}\right) \end{pmatrix}^{\alpha\beta},
\end{aligned} \tag{2.50}$$

where $\tilde{H}$ is a shorthand for the preconditioned correction to the Hamiltonian given by Eq. 2.48.

## 2.5 Variations with Respect to the NGWFs

Now that we have shown how to incorporate DFT + $U$ into density-matrix based methods which used a fixed nonorthogonal representation, we turn our attention to the outer energy-minimisation loop in ONETEP. We consider the contribution due to DFT + $U$ of the total-energy variation with respect to the expansion coefficients of the NGWF for a fixed, optimised, density kernel. The results of this section, of course, apply to any technique which optimises its representation functions for minimal energy, such as those described in Refs. [38–41].

The required derivative with respect to covariant support functions may be broken into its contributing parts as

$$\begin{aligned}\frac{\partial E_{DFT+U}}{\partial \phi_{\gamma}(\mathbf{r})} &= \frac{\partial E_{DFT+U}}{\partial \tilde{K}^{\theta\eta}} \left( \frac{\partial \tilde{K}^{\eta\theta}}{\partial K^{\iota\kappa}} \frac{\partial K^{\iota\kappa}}{\partial \phi_{\gamma}(\mathbf{r})} + \frac{\partial \tilde{K}^{\eta\theta}}{\partial S_{\lambda\nu}} \frac{\partial S_{\lambda\nu}}{\partial \phi_{\gamma}(\mathbf{r})} \right) \\ &\quad + \sum_{J} \frac{\partial E_{DFT+U}}{\partial P^{(J)}_{\xi\tau}} \frac{\partial P^{(J)}_{\xi\tau}}{\partial \phi_{\gamma}(\mathbf{r})} \\ &= H^{DFT+U}_{\theta\eta} \left( \frac{\partial \tilde{K}^{\eta\theta}}{\partial K^{\iota\kappa}} \frac{\partial K^{\iota\kappa}}{\partial S_{\lambda\nu}} + \frac{\partial \tilde{K}^{\eta\theta}}{\partial S_{\lambda\nu}} \right) \frac{\partial S_{\lambda\nu}}{\partial \phi_{\gamma}(\mathbf{r})} \\ &\quad + \sum_{J} \frac{\partial E_{DFT+U}}{\partial P^{(J)}_{\xi\tau}} \frac{\partial P^{(J)}_{\xi\tau}}{\partial \phi_{\gamma}(\mathbf{r})}. \end{aligned} \tag{2.51}$$

As for the density kernel gradient, the NGWF gradient is calculated using the purified and renormalised density kernel and so contains a preconditioning term which drives the trace of the density-matrix to the correct occupancy of the system.

The covariant metric explicitly depends on the covariant functions and, assuming real-valued NGWFs for simplicity, we find that

$$\frac{\partial S_{\lambda\nu}}{\partial \phi_{\gamma}(\mathbf{r})} = \delta^{\gamma}_{\lambda} \phi_{\nu}(\mathbf{r}) + \delta^{\gamma}_{\nu} \phi_{\lambda}(\mathbf{r}). \tag{2.52}$$

The terms in the parentheses of the second to last line of Eq. 2.52 evaluate to

$$\frac{\partial \tilde{K}^{\eta\theta}}{\partial S_{\lambda\nu}} = \frac{N}{\left(K^{\alpha\beta} S_{\beta\alpha}\right)} \left[ - \frac{K^{\eta\theta}}{\left(K^{\gamma\delta} S_{\delta\gamma}\right)} K^{\lambda\nu} \right], \tag{2.53}$$

$$\frac{\partial \tilde{K}^{\eta\theta}}{\partial K^{\epsilon\zeta}} = \frac{N}{\left(K^{\alpha\beta} S_{\beta\alpha}\right)} \left[ \delta^{\eta}_{\epsilon} \delta^{\theta}_{\zeta} - \frac{K^{\eta\theta}}{\left(K^{\gamma\delta} S_{\delta\gamma}\right)} S_{\epsilon\zeta} \right] \quad \text{and} \tag{2.54}$$

$$\frac{\partial K^{\epsilon\zeta}}{\partial S_{\lambda\nu}} = 3L^{\epsilon\lambda} L^{\nu\zeta} - 2L^{\epsilon\lambda} (LSL)^{\nu\zeta} - 2 (LSL)^{\epsilon\lambda} L^{\nu\zeta}. \tag{2.55}$$

Contraction between the DFT + $U$ term in the Hamiltonian and these terms provide a tensor $\tilde{Q}^{\bullet\bullet}$ which generates the contribution to the NGWF gradient due to mixing among the NGWFs, given by

$$\begin{aligned}\tilde{Q}^{\lambda\nu} &= H^{DFT+U}_{\theta\eta}\left(\frac{\partial\tilde{K}^{\eta\theta}}{\partial K^{\iota\kappa}}\frac{\partial K^{\iota\kappa}}{\partial S_{\lambda\nu}}+\frac{\partial\tilde{K}^{\eta\theta}}{\partial S_{\lambda\nu}}\right)\\ &= \frac{N}{\left(K^{\alpha\beta}S_{\beta\alpha}\right)}\left(3L\tilde{H}L-2L\tilde{H}LSL-2LSL\tilde{H}L-\mu^{DFT+U}K\right)^{\lambda\nu}.\end{aligned} \tag{2.56}$$

The remaining terms involve the Hubbard projections themselves, and changes beyond linear mixing of the NGWFs. We begin with the action of the DFT + $U$ contribution to the Hamiltonian on the correlated subspace projections, that is

$$\begin{aligned}\frac{\partial E_{DFT+U}}{\partial P^{(J)}_{\xi\tau}} &= \frac{\partial}{\partial P^{(J)}_{\xi\tau}}\left[\sum_I \frac{1}{2}U^{(I)\gamma}_{\alpha}\left(P^{(I)}\tilde{K}-P^{(I)}\tilde{K}P^{(I)}\tilde{K}\right)_{\gamma}^{\ \alpha}\right]\\ &= \frac{1}{2}U^{(J)\gamma}_{\alpha}\frac{\partial}{\partial P^{(J)}_{\xi\tau}}\left[\left(P^{(J)}\tilde{K}-P^{(J)}\tilde{K}P^{(J)}\tilde{K}\right)_{\gamma}^{\ \alpha}\right]\\ &= \frac{1}{2}U^{(J)\gamma}_{\alpha}\left(\delta^{\xi}_{\gamma}\tilde{K}^{\tau\alpha}-\delta^{\xi}_{\gamma}\left(\tilde{K}P^{(J)}\tilde{K}\right)^{\tau\alpha}-\left(P^{(J)}\tilde{K}\right)_{\gamma}^{\ \xi}\tilde{K}^{\tau\alpha}\right)\\ &= \frac{1}{2}\left(\tilde{K}U^{(J)}-\tilde{K}P^{(J)}\tilde{K}U^{(J)}-\tilde{K}U^{(J)}P^{(J)}\tilde{K}\right)^{\tau\xi}\\ &= \tilde{K}^{\tau\epsilon}H^{(J)DFT+U}_{\epsilon\zeta}P^{(J)\zeta\xi} = P^{(J)\tau\epsilon}H^{(J)DFT+U}_{\epsilon\zeta}\tilde{K}^{\zeta\xi},\end{aligned} \tag{2.57}$$

where $H^{(J)DFT+U}_{\alpha\beta}$ is the contribution from each site in Eq. 2.45. The Hubbard projection operators depend explicitly on the covariant NGWFs which overlap with their corresponding Hubbard projectors (or Hubbard projector duals) and this dependence may be expressed as

$$\begin{aligned}\frac{\partial P^{(J)}_{\xi\tau}}{\partial\phi_{\gamma}(\mathbf{r})} &= \frac{\partial}{\partial\phi_{\gamma}(\mathbf{r})}\left[V^{(J)}_{\xi m'}O^{(J)m'm}W^{(J)}_{m\tau}\right]\\ &= \delta^{\gamma}_{\xi}\varphi^{(J)}_{m'}(\mathbf{r})\,O^{(J)m'm}W^{(J)}_{m\tau}+V^{(J)}_{\xi m'}O^{(J)m'm}\varphi^{(J)}_{m}(\mathbf{r})\,\delta^{\gamma}_{\tau},\end{aligned} \tag{2.58}$$

where we have assumed real-valued Hubbard projectors. We may combine the latter two results to compute the remaining DFT + $U$ term in the NGWF gradient, that is for each site $J$,

$$\begin{aligned}\frac{\partial E_{DFT+U}}{\partial P^{(J)}_{\xi\tau}}\frac{\partial P^{(J)}_{\xi\tau}}{\partial\phi_{\gamma}(\mathbf{r})} &= \varphi^{(J)}_{m'}(\mathbf{r})\,O^{(J)m'm}W^{(J)}_{m\tau}P^{(J)\tau\epsilon}H^{(J)DFT+U}_{\epsilon\zeta}\tilde{K}^{\zeta\gamma}\\ &\quad+\tilde{K}^{\gamma\epsilon}H^{(J)DFT+U}_{\epsilon\zeta}P^{(J)\zeta\xi}V^{(J)}_{\xi m'}O^{(J)m'm}\varphi^{(J)}_{m}(\mathbf{r})\\ &= 2\tilde{K}^{\gamma\epsilon}H^{(J)DFT+U}_{\epsilon\zeta}P^{(J)\zeta\xi}V^{(J)}_{\xi m'}O^{(J)m'm}\varphi^{(J)}_{m}(\mathbf{r})\\ &= 2\tilde{K}^{\gamma\epsilon}\left[\hat{H}^{(J)DFT+U}\phi_{\epsilon}\right](\mathbf{r}).\end{aligned} \tag{2.59}$$

Here, due to the subspace-localised nature of the DFT + $U$ correction, only those NGWFs $\phi_\epsilon$ which explicitly overlap with the Hubbard projectors expressed on the grid $\varphi_m^{(J)}$ contribute and thus need to be summed over.

To conclude, the contravariant gradient of the DFT + $U$ energy with respect to the NGWFs is given by the expression

$$\frac{\partial E_{DFT+U}}{\partial \phi_\gamma(\mathbf{r})} = 2\left[\tilde{Q}^{\gamma\nu}\phi_\nu + \sum_J \tilde{K}^{\gamma\epsilon}\left[\hat{H}^{(J)DFT+U}\phi_\epsilon\right]\right](\mathbf{r}). \tag{2.60}$$

Since, however, we require a covariantly transforming NGWF update in order to improve upon those functions while preserving their tensorial character, the above contravariant gradient needs to be multiplied with the covariant metric tensor to give the necessary covariant DFT + $U$ NGWF gradient term,

$$g_\alpha(\mathbf{r}) = 2S_{\alpha\gamma}\left[\tilde{Q}^{\gamma\nu}\phi_\nu + \sum_J \tilde{K}^{\gamma\epsilon}\left[\hat{H}^{(J)DFT+U}\phi_\epsilon\right]\right](\mathbf{r}). \tag{2.61}$$

## 2.6 Variations with Respect to Ionic Positions

While DFT + $U$ is less commonly applied as a method to improve first-principles atomic geometries than to rectify local moments and spectra, recent success with corrected ionic structures [42–46] encourage us to think of DFT + $U$ as a true correction for ab initio energeties. A linear-scaling implementation of the DFT + $U$ force corrections may be useful, for example, in systems such as biological organometallic complexes, where GGA functionals may tend to systematically overbind ligands to transition-metal ions. As such, we have implemented the DFT + $U$ forces terms, as well as the total-energy minimisation scheme, in the ONETEP code. We shall now describe the required methodology.

We assume that the ground-state density for a given ionic configuration is located before the forces are computed, so that the total-energy is variationally minimised with respect to both the NGWF expansion coefficients and the matrix elements of the density kernel.

The DFT + $U$ correction then contributes to the ionic forces only via the spatial dependence of the Hubbard projection operators, that is for the ion labelled $j$,

$$\mathbf{F}_j = -\frac{\partial E_{DFT+U}}{\partial \mathbf{R}_j} = -\sum_J \frac{\partial E_{DFT+U}}{\partial P_{\xi\tau}^{(J)}}\frac{\partial P_{\xi\tau}^{(J)}}{\partial \mathbf{R}_j}. \tag{2.62}$$

In this expression, since the Hubbard projectors are usually considered to be associated with one atomic site only, the subspace index $J$ need only run over subspaces centred on ion $j$ only. We will henceforth suppress the summation symbol, for notational clarity, since the generalisation to multiple subspaces per ion is straightforward.

The spatial derivative of the NGWF representation of the Hubbard projection operator may be expressed as a spatial derivative of the (real-valued) covariant projectors and contravariant subspace metric tensor themselves. First, however, we define the NGWF representation of the spatial derivative of the Hubbard projectors as the three-component vector

$$\begin{aligned} \mathbf{X}^{(J)}_{\xi m} &= \langle \phi_\xi | \nabla | \varphi^{(J)}_m \rangle \\ &= \int d\mathbf{r}\; \phi_\xi (\mathbf{r}) \left[ \int d\mathbf{G}\; (-i\mathbf{G})\, e^{-i\mathbf{G}.\mathbf{r}} \varphi^{(J)}_m (\mathbf{G}) \right], \end{aligned} \tag{2.63}$$

and $\mathbf{Y}^{(J)}_{m\xi} = \mathbf{X}^{(J)}_{\xi m}$ is its transpose for each component. The required projection derivative is thus given by

$$\begin{aligned} \frac{\partial P^{(J)}_{\xi\tau}}{\partial \mathbf{R}_j} &= \left( \frac{\partial P^{(J)}_{\xi\tau}}{\partial \varphi^{(J)}_m (\mathbf{r})} \right) \frac{\partial \varphi^{(J)}_m (\mathbf{r})}{\partial \mathbf{R}_j} + \left( \frac{\partial P^{(J)}_{\xi\tau}}{\partial O^{(J)mm'}} \right) \frac{\partial O^{(J)mm'}}{\partial \mathbf{R}_j} \\ &= \int \left( \frac{\partial}{\partial \varphi^{(J)}_m (\mathbf{r})} \left[ V^{(J)}_{\xi m'} O^{(J)m'm''} W^{(J)}_{m''\tau} \right] \right) \\ &\quad \times \frac{\partial}{\partial \mathbf{R}_j} \left[ \int d\mathbf{G}\; e^{-i\mathbf{G}.\mathbf{r}} \varphi^{(J)}_m (\mathbf{G}) \right] d\mathbf{r} \\ &= \int \left( \phi_\xi (\mathbf{r})\, O^{(J)mm''} W^{(J)}_{m''\tau} + V^{(J)}_{\xi m'} O^{(J)m'm} \phi_\tau (\mathbf{r}) \right) \\ &\quad \times \left[ \int d\mathbf{G}\; (-i\mathbf{G})\, e^{-i\mathbf{G}.\mathbf{r}} \varphi^{(J)}_m (\mathbf{G}) \right] d\mathbf{r} \\ &= \mathbf{X}^{(J)}_{\xi m} O^{(J)mm''} W^{(J)}_{m''\tau} + V^{(J)}_{\xi m'} O^{(J)m'm} \mathbf{Y}^{(J)}_{m\tau}. \end{aligned} \tag{2.64}$$

Here, we have neglected the term involving the partial derivative of the metric tensor with respect to ionic positions, as it vanishes for the localised subspace representation we use here and which we go on to discuss in more detail in Chap. 4. The neglect of this term is appropriate when using either conventional atom-centred system-independent Hubbard projectors or the self-consistently determined variety which we describe in Chap. 3 and Ref. [25], since a rigid displacement of all Hubbard projectors for a correlated subspace does not change the metric tensor on that subspace.

Combining this expression with the result of Eq. 2.57, in order to evaluate the force expression of Eq. 2.62, we conclude that the tensorially consistent DFT + $U$ contribution to the ionic forces is given, again simplifying using the real-valued nature of both Hubbard projectors and NGWFs, by the easily-evaluated trace of sparse matrices

$$\mathbf{F}_j = -2\mathbf{X}^{(J)}_{\xi m} O^{(J)mm''} W^{(J)}_{m''\tau} H^{(J)DFT+U}_{\tau\zeta} \tilde{K}^{\zeta\xi}. \tag{2.65}$$

## 2.7 Scaling Tests on Nickel Oxide Nano-Clusters

The first row transition-metal monoxide NiO has, for some considerable time, posed difficulties to Kohn–Sham density-functional theory and to electronic structure theories generally. As such, it has served as a valuable proving-ground for novel approaches such as periodic unrestricted Hartree-Fock theory [47], the self-interaction corrected local density approximation [19], the *GW* approximation [48], LDA + DMFT [49] and first-principles methods for calculating the Hubbard $U$ parameter [16, 50] in DFT + $U$.

Experimentally, the paramagnetic phase of NiO is found to possess a rock-salt crystal structure with a lattice constant of approximately 4.17 Å [10]. At ambient temperature, NiO is a type-II antiferromagnetic insulator with a local magnetic moment of between 1.64 and 1.9 $\mu_B$ [16], with a Néel temperature of approximately 523 K [9].

Due to the persistence of the magnetic moment and the optical gap, which lies at approximately 4 eV, of this material above the Néel temperature, it falls broadly into the category of a Mott insulator [47]. However, the material has been somewhat reclassified as a charge-transfer insulator since experimental data has shown that the states close to the top of the valence band possess a predominantly oxygen $2p$ character while those in the conduction band are of nickel $3d$-orbital character [51].

Irrespective of nomenclature, it has long been recognised that LDA-type exchange correlation functionals [8] qualitatively fail to reproduce the physics of this material, grossly under-estimating the local magnetic moment, the Kohn–Sham gap (if it is imbued with a physical interpretation) and assigning an incorrect fully $3d$-orbital character to the valence band edge, but that the DFT + $U$ method successfully corrects these deficiencies [10, 16, 21, 50, 52].

### *2.7.1 Computational Methodology*

We performed scaling tests on NiO nano-clusters of varying size, comparing the computational effort required for DFT + $U$ and uncorrected DFT calculations. We have chosen approximately spherical nano-clusters with an even number of nickel ions, so that we may tentatively assume an open-shell singlet multiplicity, analogous to the bulk antiferromagnetic ground state. In fact, however, we might expect that a transition to a ferrimagnetic or ferromagnetic ground state occurs below some critical cluster size, as it has been predicted for very small iron oxide clusters of interest for data-storage technology [53, 54].

While it is certainly of worthwhile to explore this possibility further using linear-scaling DFT + $U$, it exceeded the scope of this study, since the spin multiplicity has no direct bearing on computational expense. Moreover, since calculations on nano-clusters of varying sizes may be expected to exhibit differing convergence behaviour, we simply ran the energy-minimisation algorithm for a fixed number of iterations and did not attempt to achieve the ground-state. In Chap. 4, published in Ref. [27], we return to this material in its bulk form, addressing the DFT + *U* description of its physical properties in detail.

### *2.7.2 Scaling of Computational Effort for DFT + U*

A moderate set of run-time parameters was used, since our principal interest was to test the increase in effort needed when the DFT + *U* functionality was included in the calculation. These parameters included a 500 eV equivalent plane-wave cutoff energy, a spin-polarised density kernel, the LSDA XC functional [8], a 25 $a_0$ density kernel cutoff with 7.5 $a_0$ NGWF cutoff radii, nine NGWFs for each nickel ion and 4 each for oxygen. A fixed number of one NGWF update step and three density kernel update steps, with three penalty-functional idempotency corrections per density kernel update step, were used to test the scaling behaviour without the dependence of convergence behaviour on system size. NGWF overlap matrix inversion was carried out using Hotelling's algorithm [55] and a cubic simulation super-cell of three times the diameter of each nano-cluster was used, up to a maximum super-cell size of approximately 300 $a_0$. Hydrogenic projectors of the type discussed in Chap. 3 were used for DFT + *U*.

Figure 2.2 shows computational timing data for ONETEP energy-minimisation of selected NiO nano-clusters containing up to 7,153 atoms on 300 nodes of a commodity supercomputer. A reasonable linear fit was obtained in spite of the rather small number of data points available; the available memory was exceeded when attempting calculations of a larger cluster of 11,513 atoms. The zero-time intercept lay at 450–500 atoms, indicating very efficient initialisation of the pre-requisite data in these calculations.

The NiO nano-clusters are by no means a favourable case for the DFT + *U* method, since approximately half of the ions host correlated subspaces. Nonetheless, we see a very small increase in computational time when the DFT + *U* functionality is applied, of approximately 5%, and preservation of linear-scaling performance.

Figure 2.3 shows the time spent computing the DFT + *U* projection operator and its contribution to the total-energy and Hamiltonian. This indicates that no aspect of this functionality appreciably deviates from linear-scaling behaviour. Also shown is the timing for one calculation of the DFT + *U* contribution to the ionic forces, also exhibiting favourable scaling for those calculations which fell within memory resources. Significantly, however, we note that the total time spent in these subroutines makes up only a small fraction of the increase in cost incurred by DFT + *U*, remaining at less than 1% of the total computational time.

In order to understand where the dominant contribution to the additional cost originates, if not in the DFT + *U* subroutines themselves, we direct the reader to Fig. 2.4, where the size-dependent sparsity of some important matrices, with and without DFT + *U*, is quantified.

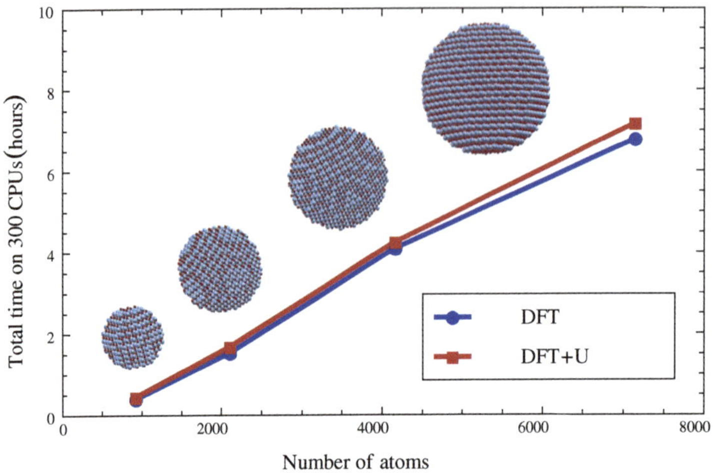


**Fig. 2.2** Scaling tests of energy-minimisation functionality, including three density kernel optimisation steps and one NGWF update step, comparing DFT and DFT + $U$. Four sizeable nano-clusters of NiO were tested on 300 processing cores

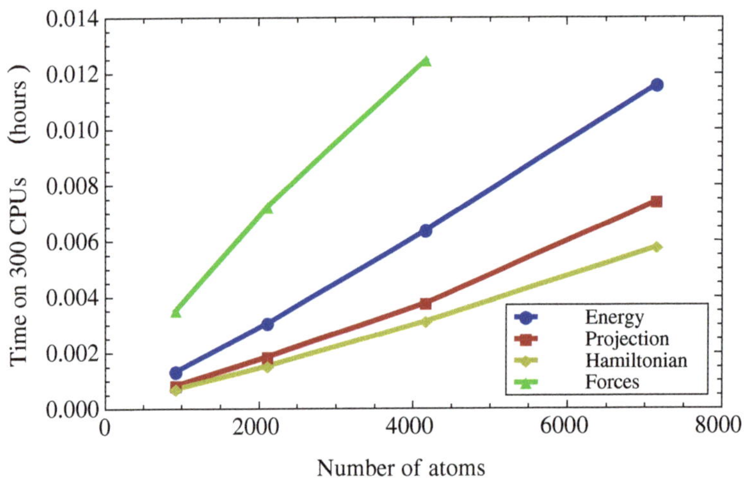


**Fig. 2.3** Timing of the DFT + $U$ subroutines in the test calculations shown in Fig. 2.2

For a conventional DFT calculation, the sparsity of the Hamiltonian matrix is dominated by the NGWF representation of the non-local pseudopotential. This in turn is computed using the product of the overlap matrix between the NGWFs and the non-local pseudopotential projectors with its transpose. In essence, pairs of NGWFs which overlap with a common non-local projector must be represented in the Hamiltonian, and the same is true of pairs of NGWFs overlapping with a common Hubbard projector when DFT + $U$ is used.

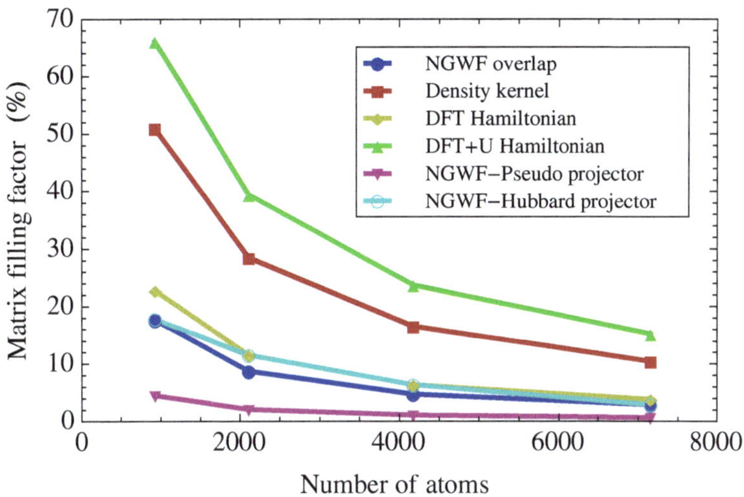


**Fig. 2.4** Nano-cluster size dependence of filling factors of principal matrices. These matrix fillings partially determine the computational cost for DFT and DFT + $U$

Non-local pseudopotential projectors tend to have radii not in excess of 2 $a_0$ for lighter elements, up to and including the first-row transition metals. However, the Hubbard projectors may require radii significantly greater than this, as indicated by the Hubbard projector density shown in Fig. 3.1 for 3$d$-type projectors.

In our implementation of DFT + $U$, we attributed a cutoff radius to all Hubbard projectors equal to the NGWF cutoff radii of their host ions, in this case 7.5 $a_0$. This was primarily to allow the use of self-consistently determined Hubbard projectors in the form of NGWFs, as we go on to discuss in Chap. 3. When using hydrogenic projectors of smaller characteristic radii, it is almost certainly sufficient to use reduced Hubbard projector cutoff radii for computational efficiency, but we did not explore this possibility. The significant increase in the filling of the Hamiltonian matrix in DFT + $U$ over DFT, and thus the computation of it products with other quantities such as the density kernel, is thus largely responsible for the incurred increase in computational expense. The increased Hamiltonian filling has consequences too for the calculation of the gradient of the energy with respect to the NGWF expansion coefficients, as indicated in Fig. 2.5, which shows the fractional increase in time spent in some principal operations in the energy-minimisation algorithm.

Due to the increase in the number of matrix elements in the Hamiltonian when the DFT + $U$ contribution is included, it takes close to twice as much effort to calculate its expansion on the *psinc* grid. Moreover, since the grid-expansion of the action of the Hamiltonian on the NGWFs is also required for the energy gradient with respect to NGWFs, this too is made more costly by DFT + $U$.

Of course, when using hydrogenic Hubbard projectors, at least for first-row transition metal ions, we could safely reduce the cutoff radii of the Hubbard projectors and

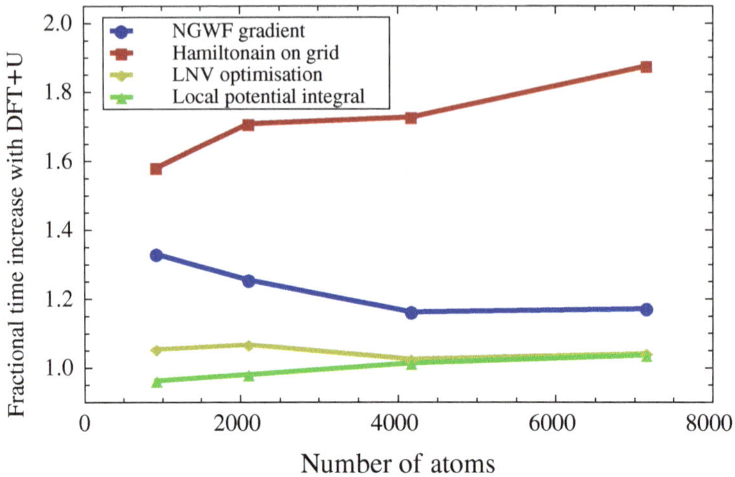


**Fig. 2.5** Fractional increase in time spent on selected operations when the DFT + $U$ functionality is activated

therefore the number of matrix elements included in the Hamiltonian, if so desired. If carried out, this would be expected to further reduce the increase in linear-scaling pre-factor for these systems from approximately 5% closer to the much lower fraction which is purely due to the DFT + $U$ subroutines.

## 2.8 Concluding Remarks

We have detailed a linear-scaling implementation of the widely-used DFT + $U$ method for treating strongly-correlated systems from first-principles. The formalism is generally appropriate to methods which minimise the energy with respect to the single-particle density-matrix, and allows for the optimisation of a nonorthogonal representation, nonorthogonal Hubbard projectors and ionic positions.

We have demonstrated the preservation of linear-scaling performance on strongly-correlated nano-clusters in excess of 7,000 atoms. Even for systems such as these, with a high density of correlated sites, the increase in computational pre-factor is rather modest. The method is, furthermore, expected to incur negligible extra cost in large systems comprising only a small number of Hubbard subspaces.

We expect that our method may be helpful in bringing linear-scaling DFT to bear on more problematic systems than those to which it is usually applied, for example binding-sites in organometallic enzymes, heterostructures containing magnetic layers for data storage and processing, defective oxides and interfaces with catalytic oxide surfaces.

# References


1. J.A. Pople, R.K. Nesbet, Self-consistent orbitals for radicals. J. Chem. Phys. **22**, 572 (1954)
2. P. Hohenberg, W. Kohn, Inhomogeneous electron gas. Phys. Rev. **136**(3B), B864 (1964)
3. W. Kohn, L.J. Sham, Self-consistent equations including exchange and correlation effects. Phys. Rev. **140**(4A), A1133 (1965)
4. J.S. Miller, A.J. Epstein, Organic and organometalling molecular magnetic materials—designer magnets. Angew. Chem. Int. Ed. Engl. **33**, 385 (1994)
5. B.C.H. Steele, A. Heinzel, Materials for fuel-cell technologoes. Nature **414**, 345 (2001)
6. R.H. Holm, P. Kennepohl, E.I. Solomon, Structural and functional aspects of metal sites in biology. Chem. Rev. **96**, 2239 (1996)
7. N.F. Mott, The basis of the electron theory of metals, with special reference to the transition metals. Proc. Phys. Soc. A **62**, 416 (1949)
8. J.P. Perdew, A. Zunger, Self-interaction correction to density-functional approximations for many-electron systems. Phys. Rev. B **23**(10), 5048 (1981)
9. K. Terakura, T. Oguchi, A.R. Williams, J. Kübler, Band theory of insulating transition-metal monoxides: Band-structure calculations. Phys. Rev. B **30**(8), 4734 (1984)
10. S.L. Dudarev, G.A. Botton, S.Y. Savrasov, C.J. Humphreys, A.P. Sutton, Electron-energy-loss spectra and the structural stability of nickel oxide: an LSDA+U study. Phys. Rev. B **57**(3), 1505 (1998)
11. J. Hubbard, Electron correlations in narrow energy bands. Proc. R. Soc. London Ser. A **276**, 238 (1963)
12. J. Hubbard, Electron correlations in narrow energy bands II: the degenerate band case. Proc. R. Soc. London Ser. A **277**, 237 (1964)
13. J. Hubbard, Electron correlations in narrow energy bands III: an improved solution. Proc. R. Soc. London Ser. A **281**, 401 (1964)
14. E. Artacho, del L.M. Bosch, Nonorthogonal basis sets in quantum mechanics: representations and second quantization. Phys. Rev. A **43**(11), 5770 (1991)
15. J.P. Perdew, R.G. Parr, M. Levy, J.L. Balduz, Density-functional theory for fractional particle number: derivative discontinuities of the energy. Phys. Rev. Lett. **49**(23), 1691 (1982)
16. M. Cococcioni, de S. Gironcoli, Linear response approach to the calculation of the effective interaction parameters in the LDA + U method. Phys. Rev. B **71**(3), 035105 (2005)
17. H.J. Kulik, M. Cococcioni, D.A. Scherlis, N. Marzari, Density functional theory in transition-metal chemistry: a self-consistent Hubbard U approach. Phys. Rev. Lett. **97**(10), 103001 (2006)
18. A.J. Cohen, P. Mori-Sanchez, W. Yang, Insights into current limitations of density functional theory. Science **321**(5890), 792 (2008)
19. A. Svane, O. Gunnarsson, Transition-metal oxides in the self-interaction-corrected density-functional formalism. Phys. Rev. Lett. **65**(9), 1148 (1990)
20. V.I. Anisimov, J. Zaanen, O.K. Andersen, Band theory and Mott insulators: Hubbard U instead of Stoner I. Phys. Rev. B **44**(3), 943 (1991)
21. V.I. Anisimov, I.V. Solovyev, M.A. Korotin, M.T. Czyżyk, G.A. Sawatzky, Density-functional theory and NiO photoemission spectra. Phys. Rev. B **48**(23), 16929 (1993)
22. V.I. Anisimov, A.I. Poteryaev, M.A. Korotin, A.O. Anokhin, G. Kotliar, First-principles calculations of the electronic structure and spectra of strongly correlated systems: dynamical mean-field theory. J. Phys. Condens. Matt. **9**(35), 7359 (1997)
23. A.I. Lichtenstein, M.I. Katsnelson, Ab initio calculations of quasiparticle band structure in correlated systems: LDA++ approach. Phys. Rev. B **57**(12), 6884 (1998)
24. S. Atwell, E. Meggers, G. Spraggon, P.G. Schultz, Structure of a copper-mediated base pair in DNA. J. Am. Chem. Soc. **123**(49), 12364 (2001)
25. D.D. O'Regan, N.D.M. Hine, M.C. Payne, A.A. Mostofi, Projector self-consistent DFT+U using nonorthogonal generalized Wannier functions. Phys. Rev. B **82**(8), 081102 (2010)

26. L.V. Pourovskii, B. Amadon, S. Biermann, A. Georges, Self-consistency over the charge density in dynamical mean-field theory: a linear muffin-tin implementation and some physical implications. Phys. Rev. B **76**(23), 235101 (2007)
27. D.D. O'Regan, M.C. Payne, A.A. Mostofi, Subspace representations in ab initio methods for strongly correlated systems. Phys. Rev. B **83**(24), 245124 (2011)
28. V.I. Anisimov, F. Aryasetiawan, A.I. Lichtenstein, First-principles calculations of the electronic structure and spectra of strongly correlated systems: the LDA + U method. J. Phys. Condens. Matt. **9**(4), 767 (1997)
29. A.I. Liechtenstein, V.I. Anisimov, J. Zaanen, Density-functional theory and strong interactions: orbital ordering in Mott–Hubbard insulators. Phys. Rev. B **52**(8), R5467 (1995)
30. M.T. Czyżyk, G.A. Sawatzky, Local-density functional and on-site correlations: the electronic structure of $La_2CuO_4$ and $LaCuO_3$. Phys. Rev. B **49**(20), 14211 (1994)
31. D.-K. Seo, Self-interaction correction in the LDA + U method. Phys. Rev. B **76**, 033102 (2007)
32. R. McWeeny, Some recent advances in density matrix theory. Rev. Mod. Phys. **32**(2), 335 (1960)
33. X.-P. Li, R.W. Nunes, D. Vanderbilt, Density-matrix electronic-structure method with linear system-size scaling. Phys. Rev. B **47**(16), 10891 (1993)
34. R.W. Nunes, D. Vanderbilt, Generalization of the density-matrix method to a nonorthogonal basis. Phys. Rev. B **50**(23), 17611 (1994)
35. M.S. Daw, Model for energetics of solids based on the density matrix. Phys. Rev. B **47**(16), 10895 (1993)
36. P.D. Haynes, C.-K. Skylaris, A.A. Mostofi, M.C. Payne, Density kernel optimization in the ONETEP code. J. Phys. Condens. Matt. **20**(29), 294207 (2008)
37. C.A. White, P. Maslen, M.S. Lee, M. Head-Gordon, The tensor properties of energy gradients within a non-orthogonal basis. Chem. Phys. Lett. **276**(1–2), 133 (1997)
38. F. Mauri, G. Galli, R. Car, Orbital formulation for electronic-structure calculations with linear system-size scaling. Phys. Rev. B **47**(15), 9973 (1993)
39. F. Mauri, G. Galli, Electronic-structure calculations and molecular-dynamics simulations with linear system-size scaling. Phys. Rev. B **50**(7), 4316 (1994)
40. J. Kim, F. Mauri, G. Galli, Total-energy global optimizations using nonorthogonal localized orbitals. Phys. Rev. B **52**(3), 1640 (1995)
41. P. Ordejón, D.A. Drabold, R.M. Martin, M.P. Grumbach, Linear system-size scaling methods for electronic-structure calculations. Phys. Rev. B **51**(3), 1456 (1995)
42. D.A. Scherlis, M. Cococcioni, P. Sit, N. Marzari, Simulation of heme using DFT + U: a step toward accurate spin-state energetics. J. Phys. Chem. B **111**(25), 7384 (2007)
43. H.J. Kulik, N. Marzari, A self-consistent Hubbard U density-functional theory approach to the addition–elimination reactions of hydrocarbons on bare $FeO^+$. J. Chem. Phys. **129**(13), 134314 (2008)
44. H.J. Kulik, N. Marzari, Systematic study of first-row transition-metal diatomic molecules: a self-consistent DFT + U approach. J. Chem. Phy. **133**(11), 114103 (2010)
45. H.J. Kulik, L.C. Blasiak, N. Marzari, C.L. Drennan, First-principles study of non-heme Fe(II) halogenase SyrB2 reactivity. J. Am. Chem. Soc. **131**(40), 14426 (2009)
46. H. Hsu, K. Umemoto, M. Cococcioni, R. Wentzcovitch, First-principles study for low-spin $LaCoO_3$ with a structurally consistent Hubbard U. Phys. Rev. B **79**(12), 125124 (2009)
47. M.D. Towler, N.L. Allan, N.M. Harrison, V.R. Saunders, W.C. Mackrodt, E. Aprà, Ab initio study of MnO and NiO. Phys. Rev. B **50**(8), 5041 (1994)
48. F. Aryasetiawan, O. Gunnarsson, Electronic structure of NiO in the GW approximation. Phys. Rev. Lett. **74**(16), 3221 (1995)
49. X. Ren, I. Leonov, G. Keller, M. Kollar, I. Nekrasov, D. Vollhardt, LDA + DMFT computation of the electronic spectrum of NiO. Phys. Rev. B **74**(19), 195114 (2006)
50. W.E. Pickett, S.C. Erwin, E.C. Ethridge, Reformulation of the LDA + U method for a local-orbital basis. Phys. Rev. B **58**(3), 1201 (1998)

51. G.A. Sawatzky, J.W. Allen, Magnitude and origin of the band gap in NiO. Phys. Rev. Lett. **53**(24), 2339 (1984)
52. O. Bengone, M. Alouani, P. Blöchl, J. Hugel, Implementation of the projector augmented-wave LDA+U method: application to the electronic structure of NiO. Phys. Rev. B **62**(24), 16392 (2000)
53. S. López, A.H. Romero, J. Mejía-López, J. Mazo-Zuluaga, J. Restrepo, Structure and electronic properties of iron oxide clusters: a first-principles study. Phys. Rev. B **80**(8), 085107 (2009)
54. K. Palotás, A.N. Andriotis, A. Lappas, Structural, electronic, and magnetic properties of nanometer-sized iron-oxide atomic clusters: comparison between GGA and GGA+U approaches. Phys. Rev. B **81**(7), 075403 (2010)
55. T. Ozaki, Efficient recursion method for inverting an overlap matrix. Phys. Rev. B **64**(19), 195110 (2001)

# Chapter 3
# Projector Self-Consistent DFT+U Using Nonorthogonal Generalised Wannier Functions

The DFT+$U$ energy functional depends not only on the electronic density and the Hubbard interaction parameters, but additionally on a set of projections which delineate the correlated subspaces.

In this chapter, we describe a formulation of the density-functional theory + Hubbard model (DFT+$U$) method that is self-consistent over the choice of projectors used to define these subspaces. In order to overcome the arbitrariness in this choice, we propose the use of nonorthogonal generalised Wannier functions (NGWFs) as projectors for the DFT+$U$ correction. We iteratively refine these NGWF projectors and, hence, the DFT+$U$ functional, such that the correlated subspaces are fully self-consistent with the DFT+$U$ ground-state.

We discuss the convergence characteristics of this algorithm and compare some ground-state properties thus computed with those calculated using hydrogenic projectors. The prescribed approach is implemented within, but not restricted to, a linear-scaling DFT framework, namely the ONETEP code, and may be applied to any optimisation procedure for localised orbitals. An abbreviated form of this chapter has been published in Ref. [1]. Reprinted with permission from David D. O'Regan, Nicholas D.M. Hine, Mike C. Payne and Arash A. Mostofi, Phys. Rev. B 82, 081102(R) (2010). Copyright (2010) by the American Physical Society.

## 3.1 Localised Strongly-Correlated Subspaces

Localised correlated subspaces, central to strongly correlated ab initio methods such as DFT + Hubbard $U$ (DFT+$U$) [2, 3] and DFT + dynamical mean field theory (DFT+DMFT) [4, 5], are usually defined by a set of $3d$ and/or $4f$ atomic-like orbitals, termed "*Hubbard projectors*", that must, in general, be chosen a priori. Any physically plausible set of localised functions may, in principle, be used as Hubbard projectors and those that have been previously employed include hydrogenic wavefunctions [6, 7], Maximally-localised Wannier functions [8], and LMTO-type atomic orbitals [2, 9]. The choice of such projections introduces an adjustable parameter (in the generalised sense of a vector field) which, as we go on to show for the DFT+$U$

D. D. O'Regan, *Optimised Projections for the Ab Initio Simulation of Large and Strongly Correlated Systems*, Springer Theses, DOI: 10.1007/978-3-642-23238-1_3,

## 3.3 The Spatial Form of Hydrogenic Subspaces

Solutions of appropriate orbital symmetry of the Schrödinger equation for the hydrogenic-like atom, such as linear muffin-tin orbitals, are a common choice of Hubbard projectors [2, 9, 13] for DFT+$U$ calculations. These hydrogenic projectors are generally characterised by an effective charge $Z$ and effective mass ratio $a_\mu$, the ratio of which in turn determines their spatial diffuseness.

Of course, for a given value of $U$, results of DFT+$U$ calculations with different values chosen for $Z$ will not necessarily yield the same ground-state properties [13, 16]. Let us briefly discuss the properties of such hydrogenic orbitals, which we use in this study to investigate the dependence of computed DFT+$U$ observables on the choice of spherically-symmetric projectors. We also propose a simple method for estimating an appropriate $Z$ parameter appropriate to pseudopotential methods.

Hydrogenic projectors are defined on a radial real-space grid according to the familiar formula for the nodeless $l = n - 1$ solutions of the hydrogen atom

$$\varphi_{n,l=n-1}(r) = N \left( \frac{Z}{n a_\mu} \right)^{\frac{3}{2}} \left( \frac{Zr}{a_\mu} \right)^{n-1} \exp\left( -\frac{Zr}{n a_\mu} \right), \tag{3.4}$$

where the constant $N$ provides the appropriate normalisation as per

$$\int_0^\infty |\varphi(r)|^2 r^2 dr = 1. \tag{3.5}$$

Equivalently, and often more conveniently for computational purposes, the Fourier-Bessel transform [17],

$$\varphi_{n,l=n-1}(q) = \sqrt{\frac{2}{\pi}} \int_0^\infty \varphi_{n,l=n-1}(r)\ j_l\ (gr) r^2 dr, \tag{3.6}$$

where $j_l$ is the spherical Bessel function of the first kind, may be used to construct the normalised nodeless projectors, on the reciprocal-space radial grid, of the general form

$$\varphi_{n,l=n-1}(q) = \frac{M q^{n-1} Z^{n+\frac{3}{2}}}{\left(Z^2 + (nq)^2\right)^{n+1}}; \tag{3.7}$$

the normalisation constant $M$ being related to $N$. Shown in Figs. 3.1 and 3.2 are the probability densities of $3d$ orbital projectors, in real and reciprocal space, respectively, for various values of the effective nuclear charge $Z$. The ratio of effective masses is hereafter absorbed into the effective $Z$ for convenience.

In real multi-atomic systems, the $3d$ manifold is always somewhat distorted from spherical symmetry by hybridisation effects. This effect is expected to be more

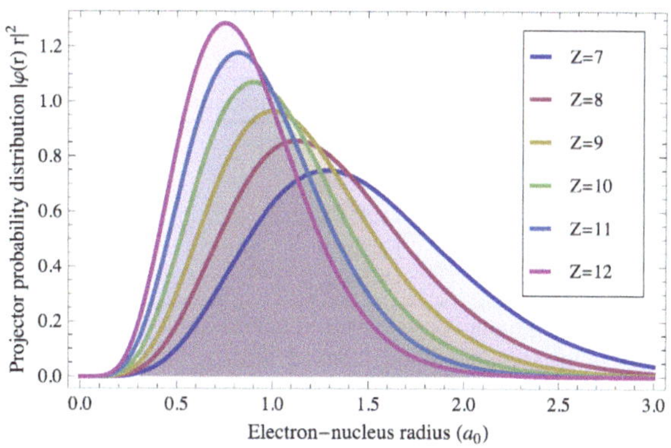


**Fig. 3.1** Normalised radial probability distribution on the real-space radial grid for the hydrogenic Hubbard projectors of 3*d* character, plotted for various values of the effective nuclear charge *Z*

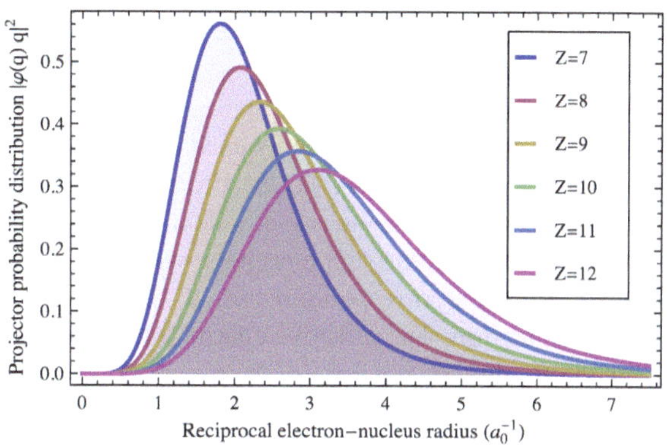


**Fig. 3.2** Normalised radial probability distribution on the reciprocal-space radial grid for the hydrogenic Hubbard projectors of 3*d* character, plotted for various values of the effective nuclear charge *Z*

pronounced in the case of molecular systems, since in this case the orbitals are more spatially delocalised than their counterparts in solids. Thus, constraining the form of the Kohn–Sham orbitals via the DFT+*U* with hydrogenic Hubbard projectors may be limited, in some cases, in its capacity to recover the subtleties of the localised electronic structure which are often responsible for the interesting physical behaviour in transition-metal containing clusters, surfaces or defective solids.

For example, in the case of binding of small molecules to catalytic solid-oxide surfaces, where the DFT+*U* correction is often beneficial, spurious Hubbard occupancies have been suspected of leading to inaccurate predictions of binding energies [16]. Additionally, even in pristine solids, the spherical-symmetry approximation may be inappropriate in cases where the correlated states are strongly hybridised with those of greater band-dispersion (which are usually associated with neighbouring ligands) and so the corresponding orbitals may differ significantly from those representable by atomic wave-functions.

Nonetheless, the use of the valence pseudo-orbitals of appropriate orbital character as Hubbard projectors is, however, somewhat justifiable as a first approximation, since these are consistent with the pseudised isolated-atom core potential which is not vastly dissimilar to that felt by the spatially localised Kohn–Sham states.

**Table 3.1** Effective nuclear charges for 3*d*-orbitals of the late transition-metal ions calculated using various methods

| Species | Clementi–Raimondi | Pseudo max. | Pseudo RMS |
|---|---|---|---|
| Manganese | 10.5282 | 9.0188 | 9.3329 |
| Iron | 11.1798 | 8.7957 | 8.4575 |
| Cobalt | 11.8554 | 8.8932 | 8.7351 |
| Nickel | 12.5295 | 9.0242 | 9.0251 |
| Copper | 13.2006 | 9.0636 | 9.2514 |

Clementi–Raimondi corresponds to pioneering atomic Hartree–Fock calculations described in Ref. [20]. In the second column, labelled "Pseudo Max." are those values for $Z$ computed by fitting the maximum radial probabilities of relativistically and non-linear core corrected PBE valence atomic pseudo-orbitals for these species to the analytical form for hydrogenic orbitals. The third column, "Pseudo RMS" is as in the previous, but the $Z$ is computed by minimising the RMS deviation (integrated up to a radius of $10\,a_0$) between the hydrogenic pseudo-orbitals and their hydrogenic counterparts

This technique is most commonly used in plane-wave pseudopotential methods such as CASTEP [18] and QUANTUM-ESPRESSO [19].

In order to simulate the use of valence pseudo-orbitals as Hubbard projectors in ONETEP, we use hydrogenic orbitals with a radial probability distribution which is fitted to that of the corresponding pseudo-orbitals for the precise pseudopotential used in the calculation. The effective charge felt by the valence electrons of 3*d* character depends not only on the form of the pseudopotential but also on the screening effects of inner valence electrons and on the chemical environment.

Similar effective charge values were found by matching the maxima of the probability distributions (the maximum radial probability for hydrogenic orbitals of principal quantum number $n = l + 1$ is attained at $r_{max} = \frac{n^2}{Z}$) and minimising their root mean squared deviation, and generally the hydrogenic orbitals are slightly more spatially compact than the true valence pseudo-orbitals, while having a smoother profile about the peak value.

Table 3.1 shows the effective charge values for the 3*d*-orbitals in some selected transition metal species. The results for iron, a particular example where we observe that there is some noticeable deviation between the $Z$ values computed by either matching the maxima of the pseudo-orbital and hydrogenic orbital or minimising their root mean squared difference, are illustrated in Fig. 3.3. As we go on to show in a study of iron porphyrin, $Z$ values estimated using either method provide projectors which perform close to optimally with respect to the set of orbitals described by Eq. 3.4.

## 3.4 Wannier Functions for Localised Subspaces

In order to obtain accurate occupancy matrices in DFT+$U$, a set of projectors is required which adequately accounts for electronic hybridisation and which, if possible, is defined unambiguously for the system under study without the need for

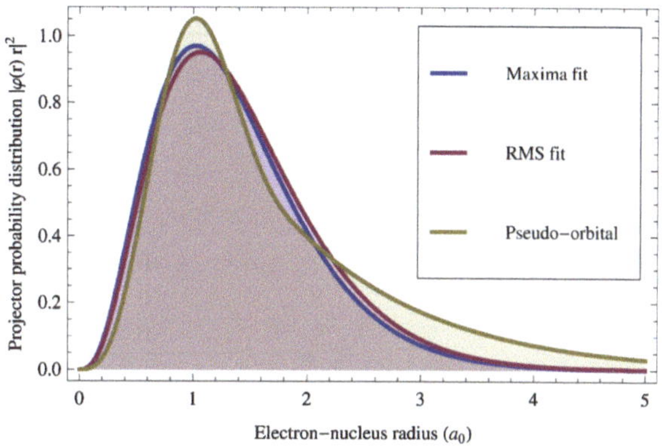


**Fig. 3.3** Normalised radial probability distribution on the real-space radial grid for the iron valence pseudo-orbitals of 3*d* character, plotted with the hydrogenic profiles fitted by matching maxima and minimising the RMS deviation

any any a priori assumptions. Wannier functions, in particular Maximally Localised Wannier functions (MLWFs) [21, 22], are computed by minimising the quadratic spread of the Wannier functions for a set of Kohn–Sham eigenstates and provide a complete and minimal basis with which to construct tight-binding models from ab initio orbitals.

Wannier functions have been used with good effect to augment DFT with localised many-body interaction corrections [23–26], in which application they exactly reproduce the nature of the chemical hybridisation of the correlated subspaces, while simultaneously retaining the crucial attribute of spatial localisation. There is, furthermore, numerical evidence to suggest that MLWFs constitute the projector set which maximises the on-site Coulomb repulsion, and hence the $U$ parameter [8]. We will return to this topic in Chap. 7.

The NGWFs generated by the energy minimisation scheme of the ONETEP linear-scaling code [27, 28] are a readily accessible set of localised orbitals which are calculated with linear-scaling computational cost.

Thus, in the ONETEP linear-scaling method, it is rather natural to use a localised subset of the NGWFs obtained at the end of a ground-state calculation, with appropriate orbital character, as Hubbard projectors for defining the DFT+$U$ occupancy matrix, since those functions are fully adapted to the chemical environment and are generated with linear-scaling cost. NGWFs, as with MLWFs, reflect the balance between the competing tendencies of electron itinerancy and localisation in strongly correlated systems and, as a result, provide an accurate representation of the occupancy of the correlated site.

The projector self-consistency technique, we go on to define, may be applied to either MLWFs or NGWFs, though we demonstrate numerical tests only for the latter. The manner in which one goes about decoupling of correlated subspaces from the remainder of the system differs somewhat depending on whether MLWFs or NGWFs are used as Hubbard projectors, and there are advantages associated with each approach. For the case of MLWFs, one is faced with the choice of whether to construct Wannier functions for the entire band-structure, and then to select the appropriate subset of these according to some criterion (e.g., localisation or angular

**Table 3.2** Spherical harmonic decomposition of the Löwdin orthonormalisation of the optimised NGWFs on a copper atom in $(CuPc)_2$ at PBE+$U$=6 eV with hydrogenic projectors, see Chap. 4 and Ref. [11]

| NGWF | s(%) | p(%) | d(%) | f(%) |
|---|---|---|---|---|
| 1 | 0.000 | 0.000 | 99.999 | 0.000 |
| 2 | 0.001 | 0.003 | 99.991 | 0.005 |
| 3 | 0.237 | 0.002 | 99.760 | 0.001 |
| 4 | 0.000 | 0.001 | 99.998 | 0.001 |
| 5 | 0.000 | 0.000 | 99.999 | 0.001 |
| 6 | 99.063 | 0.174 | 0.711 | 0.052 |
| 7 | 0.030 | 98.520 | 0.325 | 1.125 |
| 8 | 0.186 | 99.475 | 0.221 | 0.118 |
| 9 | 0.000 | 98.501 | 0.210 | 1.289 |

momentum characterisation), or to attempt to disentangle the energy bands in a user-defined energy window.

The latter method brings with it the disadvantage of necessitating user intervention in the band-selection procedure and the advantage that one then has strict control over the extent of the resulting projection over the Kohn–Sham eigenspectrum. The former method is more akin to that which is unavoidable when using NGWFs, and requires no intervention aside from the choice of a function selection procedure but in principle does not disallow some spillage of the Hubbard projection to energies outside the correlated window. NGWFs, in practice, are initialised with pure real-valued linear combinations of eigenstates of angular-momentum and tend to retain this definite character, mixing negligibly with eigenstates of different angular momentum as they undergo optimisation, as shown in Table 3.2.

## 3.5 The Self-Consistent Projector Method

Our goal in proposing a scheme whereby the Hubbard projectors are determined self-consistently is to obviate any presuppositions concerning the spatial form of the correlated subspaces. The advocated method dictates iteratively solving for the Kohn–Sham ground-state (iteratively in the literal sense that a sequence of Kohn–Sham ground-states are solved for) using as Hubbard projectors for one iteration the converged NGWFs (or indeed MLWFs, as we have discussed) from the DFT+$U$ ground-state calculation of the previous iteration.

In this way, the Hubbard projectors may converge to those that are optimally adapted for their own DFT+$U$ ground-state density. This scheme, as we go on to show, rapidly and monotonically converges to an unambiguously defined DFT+$U$ ground-state which, for a given $U$ parameter and for the systems we have studied, is of substantially lower energy than that computed using our best system-independent hydrogenic projectors (at least with the NGWF optimisation criterion of energy minimisation used here). The unambiguous definition of our procedure certainly does not imply that it is the unique projector optimisation scheme, far from it, and we would expect alternative, and perhaps physically valid, Wannierisation schemes to yield somewhat different results.

We direct the reader's attention to Fig. 3.4 for a visual representation of the projector self-consistent DFT+$U$ method. The energy minimisation scheme in ONETEP, discussed in detail in Chap. 1, considering first a conventional DFT+$U$ calculation, takes the form of two nested conjugate-gradients energy minimisation loops. In the inner loop (coloured red) the density kernel is optimised for a fixed set of NGWFs using a combination of penalty functional [29] and LNVD [30–32] methods in order to minimise the energy while maintaining the idempotency and normalisation of the density matrix. In the outer loop (coloured blue), the energy is directly minimised with respect to the expansion coefficients of the NGWFs in the psinc basis, with full variational freedom within their truncation sphere.

If a projector self-consistent DFT+$U$ calculation is required, a third, outermost, self-consistency loop (coloured green, and not an energy-minimisation cycle) is invoked in which a subset of the optimised NGWFs of appropriate orbital symmetry are selected as the Hubbard projectors and used to generate the correlated subspaces for the next total-energy minimisation. A visual illustration of such an NGWF for FeP(CO) is given in Fig. 3.5.

A set of conventional hydrogenic projectors, with a user-defined $Z$ parameter, are used for the first iteration of the scheme, and the hydrogenic projection operator is retained thereafter in order to determine which NGWFs form a Hubbard projection the product with which attains a maximal trace, and thus which should be used as projectors for the next iteration. Referring the reader once more to Table 3.2, there is no ambiguity as to which functions are appropriate for spanning the correlated subspaces for the systems we have studied.

The converged NGWFs and density kernel are carried over from one Hubbard projector iteration to initialise the next in order to expedite convergence. Sequential changes both to the total energy and to the projection over the hydrogenic initial guess, sustained over a given number of Hubbard projector iterations, are used as convergence criteria for the method.

It is important to emphasise that the energy is not directly minimised with respect to the expansion coefficients of the Hubbard projectors in the outermost loop, since it would violate the variational principle if either the Hubbard projectors or the interactions $U$ were allowed to change during energy minimisation [15].

The projector-update process alternates between direct variational minimisation of the total energy (with respect to the density) for a fixed set of projectors and renewal of the projectors (in a sense equivalent to the density-mixing method for solving non-linear systems [33, 34]—though no actual mixing with projectors from iterations is actually needed for numerical stability). In practice, the projectors rapidly converge towards those which coincide with a subset of the NGWFs which reproduce the ground state density corresponding to the DFT+$U$ correction which they themselves define.

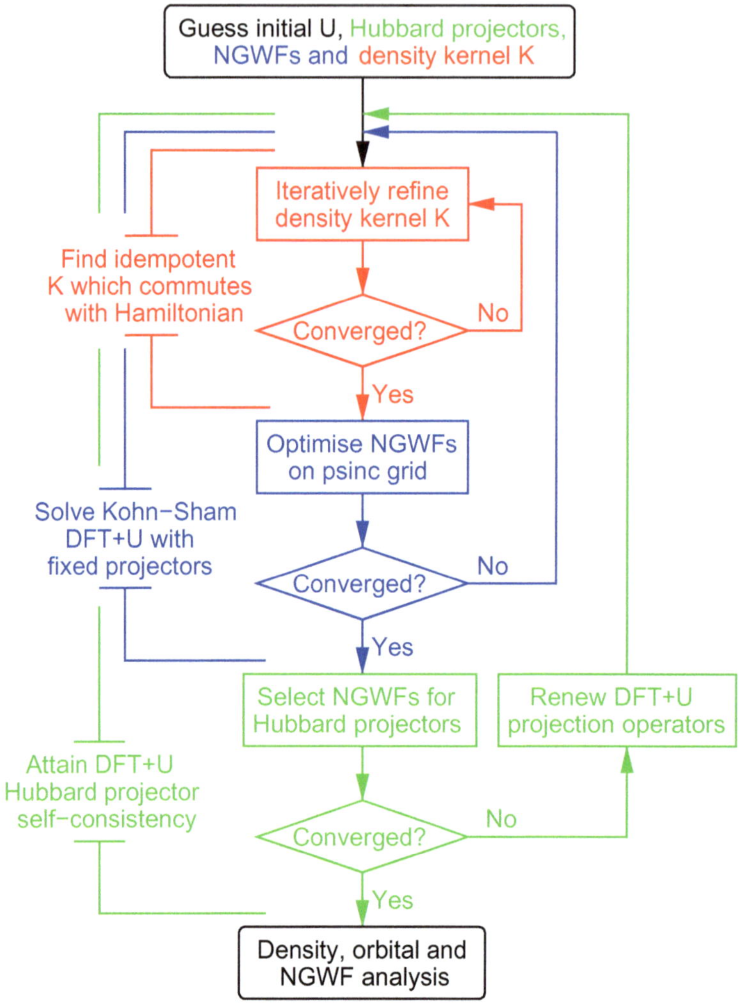


**Fig. 3.4** Schematic of the projector self-consistent DFT+$U$ scheme appropriate to the ONETEP method

## 3.6 Application to Ligated Iron Porphyrins

In order to isolate and analyse the effect of projector-dependence of DFT+$U$ ground-state properties, the effect of overlap between projections on different sites must be eliminated. The simplest way to achieve this in an unbiased fashion is, of course, to study a system where the sites do not overlap at all, or preferably one which consists of one correlated subspace in contact with a bath, thus requiring the introduction of no approximations additional to the DFT+$U$ ansatz. In this section, we study the

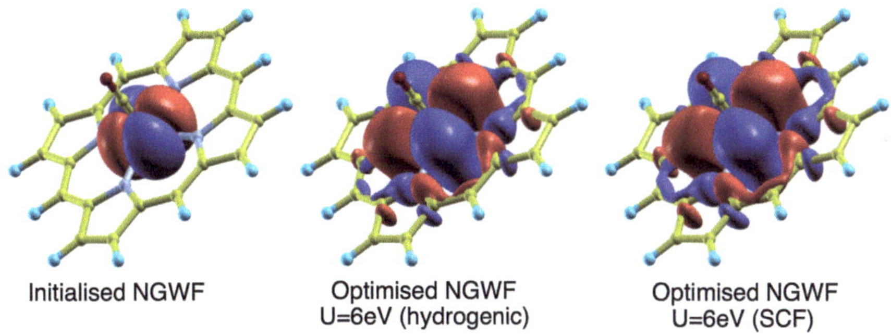


**Fig. 3.5** A localised NGWF for FeP(CO) at PBE+$U$ = 6 eV, of predominantly $3d_{xy}$ character, which is used as a Hubbard projector in the self-consistent DFT+$U$ scheme. The left image shows the initial guess for the NGWF (from the STO-3G set). The central shows the form of the NGWF after it has been fully optimised to minimise the energy in the presence of the DFT+$U$ correction (using hydrogenic Hubbard projectors with the Clementi–Raimondi effective charge [20] for iron $3d$ orbitals). The right image illustrated the final form of the NGWF after it has been refined in projector self-consistent DFT+$U$

projector-dependence of the DFT+$U$ ground state of such a single-site molecular system and apply our self-consistent approach to rectify this dependence.

### *3.6.1 Iron Porphyrin Derivatives*

We applied our proposed method to the iron porphyrin (FeP) molecule. Metalloporphyrin systems, such as FeP, play an important role in certain biochemical processes, not the least important of which involve the oxyheme-based proteins used for oxygen transport. The ability of metalloporphyrins to bind simple molecules is also of interest for technological purposes, particularly in the case of FeP which has potential for use as an electronic gas sensor since it has a greater affinity for CO and NO than $O_2$; the latter is a factor in mammalian respiration.

We have chosen the molecular systems FeP and its derivatives with axial ligands, FeP(CO), FeP($O_2$) and FeP(NO), as test cases in this study because porphyrins and metalloporphyrins have emerged as the standard test-bed for ab initio methods for treating strong correlations in molecular systems. Multiple valence character due to near-degenerate localised states on the transition-metal ions is at once the feature of metalloporphyrin complexes which makes them such effective host sites for numerous biological reactions, but also that which makes them somewhat challenging for conventional exchange-correlation functionals which tend to fractionally occupy these states.

The quantum mechanical study of iron porphyrin and its derivatives has a long and distinguished history, and is still a topic of active research. In recent years, they have been used as benchmark systems for novel methods such as DFT+$U$, see Refs

**Table 3.3** Averaged lengths of central bonds of the iron porphyrin derivative structures optimised at the PBE+$U=0$ eV level at their ground spin state, we abbreviate closed-shell singlet and open-shell singlet by CSS and OSS, respectively

| Bond length ($a_0$) | Spin-state | X–O | Fe–X | Fe–N | N–C |
|---|---|---|---|---|---|
| FeP | Triplet | – | – | 3.79 | 2.61 |
| FeP(CO) | CSS | 2.20 | 3.28 | 3.81 | 2.61 |
| FeP ($O_2$) | OSS | 2.32 | 3.38 | 3.82 | 2.61 |
| FeP(NO) | Doublet | 2.17 | 3.32 | 3.85 | 2.61 |

[35–38], many-body perturbation theory for molecular systems, as in Refs. [39, 40], and a range of calculations employing hybrid exchange-correlation functionals.

### 3.6.2 Computational Methodology

We performed fully converged energy minimisation on FeP, and its axial complexes with carbon monoxide, oxygen and nitric oxide, using the ONETEP code [27]. We used spin-polarised DFT+$U$ within the PBE generalised-gradient (GGA) [41] and pseudopotential[1] [42] approximations. An equivalent plane-wave kinetic energy cutoff of 1,000 eV was used with a cubic simulation cell of side-length 37 Å.

The NGWFs were spatially restricted to atom-centred spheres of radius 5.3 Å and no density kernel truncation was applied. Since the principal focus of this study was on the explicit dependence of computed DFT+$U$ ground-state properties on variations in the Hubbard projectors for a given $U$ value, optimised PBE ($U=0$ eV) structures were used in order to avoid any indirect effects due to the expected small resultant changes in bond lengths; details of the most crucial bond parameters are provided, with the concomitant spin multiplicity, in Table 3.3.

### 3.6.3 U and Z-Dependence of Magnetic Dipole Moments and Interaction Energies

Shown in Fig. 3.6, with data points joined by a cubic spline curve as an aid for the eye, is the interaction energy between FeP and CO as an illustration that the binding affinity between moieties in DFT+$U$ can be strongly influenced by the localisation of the Hubbard projectors. As can be seen, the binding affinity is not uniquely defined

[1] A set of RRKJ Pseudopotentials were generated using the Opium code, http://opium.sourceforge.net, using the GGA input parameters available therein, optimized for a minimum plane-wave cutoff of 680 eV, albeit with a scalar-relativistic correction for all species and, for the transition-metal ions, some slight modifications to the core radii and a non-linear core correction of Fuchs–Scheffler characteristic radius 1.3 a.u.

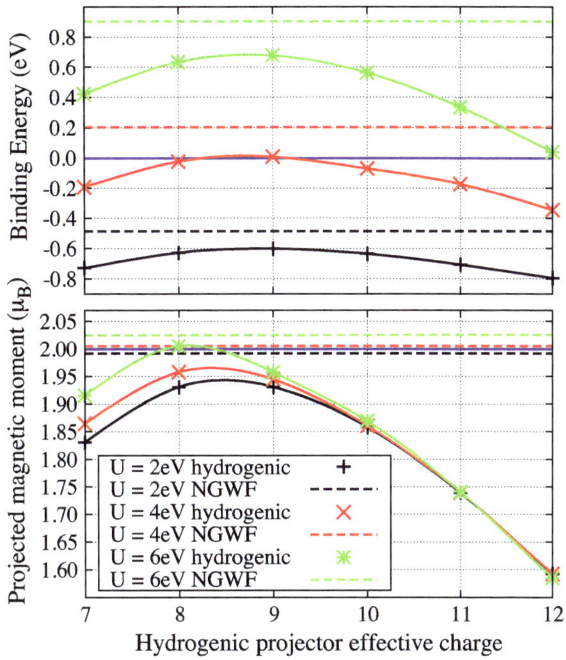


**Fig. 3.6** The interaction energy, positive for an unbound ligand, of the CO and FeP moieties (*top panel*) and the magnetic dipole moment projected onto the correlated manifold of triplet-state FeP (*bottom panel*). Both are plotted at various $U$ as a function of the effective charge $Z$ used to define the hydrogenic projectors (*solid lines*), while *dashed lines* show those quantities calculated with self-consistent NGWF Hubbard projectors. *Blue lines* indicate the binding threshold (*top*) and the ideal projected moment (*bottom*)

when hydrogenic projectors are used, although this may be partly compensated by a projector-dependent first-principles [6, 7, 13] $U$ parameter. At $U = 6$ eV it varies from approximately 0.04–0.69 eV over the range of $Z$ considered; at $U = 4$ eV the result is even qualitatively ambiguous as a function of $Z$.

Using self-consistent NGWF projectors (dashed lines) generally results in energetically less favourable ligand binding, demonstrating that, for a given value of $U$, NGWF projectors more effectively counteract the spurious tendency of GGA functionals to over-bind ligands to FeP [35]. Also shown in Fig. 3.6 is the projection of the magnetic dipole moment of FeP due to single occupancy of Kohn–Sham orbitals of primarily $d_{xz}$ and $d_{yz}$ character in its ground-state, onto the correlated subspaces. This varies strongly with the value of $Z$ chosen for the hydrogenic projectors (solid line), with only a narrow range of $Z$ at $U = 6$ eV giving values that are close to the expected 2.0 $\mu_B$ for optimal projectors.

Moreover, a pathological inconsistency with experiment emerges in that $U$ values of sufficient magnitude to achieve the requisite moment (for some $Z$) bring us into the unphysical regime where FeP+CO binding is disfavoured. Conversely, the use of self-consistent NGWF projectors (dashed) results in a projected magnetic moment which lies within the physically reasonable range and is rather insensitive to changes in $U$.

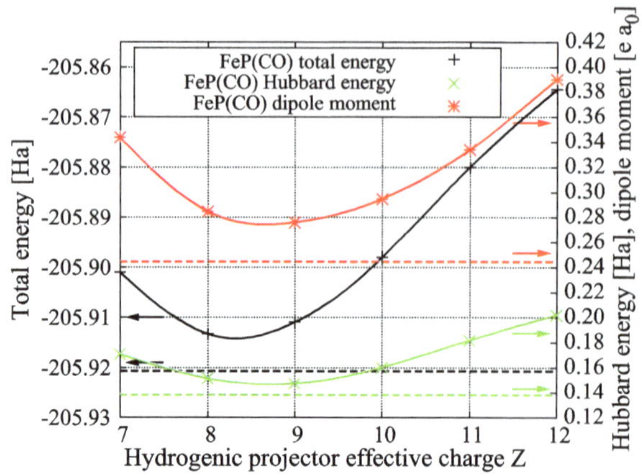


**Fig. 3.9** The total energy, DFT+$U$ contribution to the energy and the magnitude of the dipole moment of FeP(CO) at PBE+$U$=6 eV, plotted as a function of the hydrogenic projector effective charge Z. Also shown are these quantities calculated with self-consistent NGWF Hubbard projectors (*dashed lines*)

is largely due to the axial ligand binding with the transition-metal ion, takes on a minimum at a $Z$ value close to that which minimises the energy. The $Z$ in question is close to that estimated using the valence pseudo-orbitals.

We heuristically interpret the coincidence of these minima as occurring when the Hubbard projectors best match the form of the Kohn–Sham orbitals of correlated 3$d$ character, that is when those projectors are most suitable for the electronic structure of the molecule. In this situation, the correction is most efficient in reducing the spurious spatial extent of the strongly-correlated orbitals, and so their contribution to the electric dipole moment from is minimised, while at the same time minimising the DFT+$U$ contribution to the energy.

Naturally, since the diatomic ligand perturbs the iron ion's chemical environment, there is some slight difference between the $Z$ values which minimise the total-energies in FeP(CO) and FeP, see Fig. 3.10 for the latter. Changes in energy due, ostensibly, to small errors in system-independent estimates for $Z$ may readily be on the order of $10^{-2}$Ha, however, as we can see. The total-energies and electric dipole moments computed using self-consistent NGWF projectors lie, significantly, at somewhat lower values than any achievable using their hydrogenic counterparts, perhaps reflecting their greater capacity to represent subtleties in the spatial form of the strongly-correlated states.

### 3.6.7 Dependence on the Interaction Parameter U

For a given system, together with its pseudopotentials and an exchange correlation functional, the DFT+$U$ method remains dependent both on the Hubbard projectors and on the interaction parameters $U$. The projectors and parameters interdepend in a rather complicated way; on the one hand a set of Wannier function projectors may be self-consistently optimised for pre-defined $U$ parameters, on the other hand those

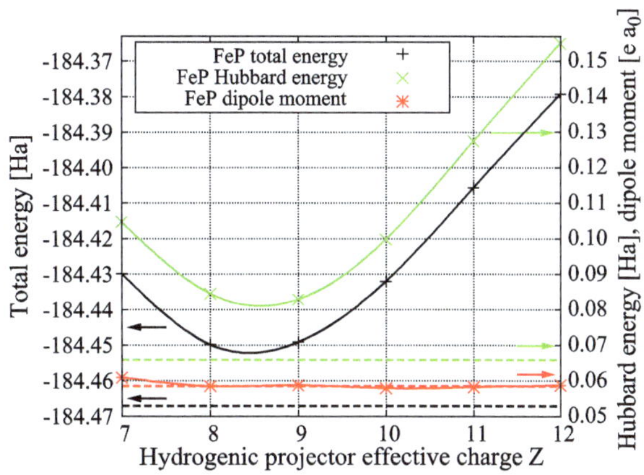


**Fig. 3.10** The total energy, DFT+$U$ contribution to the energy and the magnitude of the dipole moment of FeP at PBE+$U$=6 eV, plotted as a function of the effective charge Z used to define the hydrogenic projectors. Also shown are these quantities calculated with self-consistent NGWF Hubbard projectors (*dashed lines*)

parameters may be computed from the screened response of the system which in turn depends on the projectors. We return to this topic in in Chap. 7.

The projectors consist of a vector field at each site and contain much more information than the $U$ parameter, generally, even if the latter takes the form of a tensor. We do not subscribe to the notion, therefore, that computing the $U$ parameter may always correct for poorly-chosen Hubbard projectors.

Clearly, a self-consistency procedure over both projectors and parameters is highly desirable, and though it is not immediately clear that there may be a unique solution to such a scheme in general, this perhaps depends on how the scheme is constructed, initial conditions and of course the nature of the system it is applied to. One might, however, reasonably expect that the members of the family of solutions would yield similar physical predictions if the scheme were constructed in a physically plausible way.

While we have not undertaken such a calculations, we have investigated the $U$-dependence of various properties using both self-consistent NGWF projectors and with the conventional hydrogenic variety. The effective charge used to generate the hydrogenic orbitals was the Clementi–Raimondi charge [20] of $Z = 11.17$ for iron $3d$-orbitals. While we have already shown that this is not the $Z$ at which the energies for these species at PBE+$U$=6 eV are minimised, we chose it in order to make an unbiased comparison between the performance of the self-consistent projectors and conventional projectors constructed using a tabulated value from the literature.

Figure 3.11 shows the $U$-dependence of the total electric dipole moment in various FeP derivatives. The general effect of the self consistent NGWF projectors is to enhance the effect of the DFT+$U$ correction on the moments with respect to the hydrogenic projectors, that is to enlarge the suppression of the electric dipole moment (by localising the charge density onto the transition-metal ion and hence away from the axial ligand), and the augmentation of the magnetic dipole moment, as shown in Fig. 3.6.

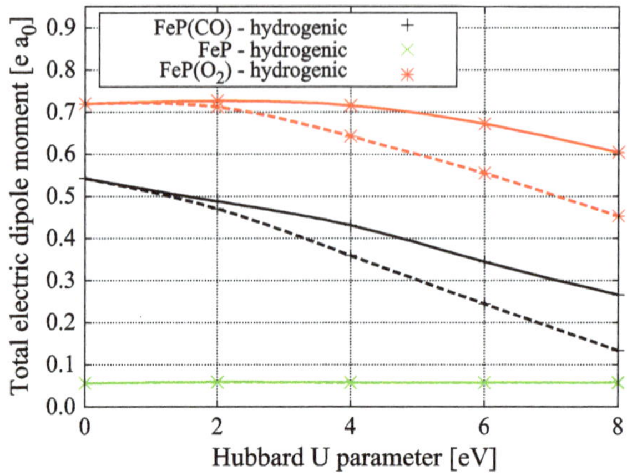


**Fig. 3.11** The magnitude of the electric dipole moment of FeP(CO) and FeP, plotted as a function of the $U$ parameter. *Solid lines* show the moments calculated using hydrogenic projectors corresponding to effective charge $Z = 11.17$. Also given are those calculated with self-consistent NGWF Hubbard projectors (*dashed lines*)

This general trend is due to the NGWFs having more weight on the axial ligands due to their ability to deviate from spherical symmetry, The very low electric dipole moment of the approximately-planar FeP molecule, incidentally, is not strongly associated with the central region of the porphyrin and so is not greatly affected by either the choice of projectors or parameter.

A complementary observable is, perhaps, the interaction energy between FeP and the diatomic axial ligands CO, $O_2$ and NO, since it is strongly dependent on both the nature of the Hubbard projectors and the $U$ parameter. This is shown in Fig. 3.12, and also in more detail for the FeP+CO case in Fig. 3.6. The observed trend in all cases is for binding to become less energetically favourable with increasing $U$, until unbinding is predicted to occur beyond a certain value which is both system dependent and, quite substantially, projector dependent.

The predicted and spurious unbinding of the ligand at higher $U$ values is due to the contraction of the $3d$-like orbitals onto the transition-metal ion and their further submergence below the valence band edge, reducing the bonding character with $\sigma$-orbitals on the axial ligand. Since these calculations were carried out at the optimised PBE+$U=0$ eV geometry, one might reasonably expect that, for both projector types, the reorganisation of the ionic geometry might go in favour of binding and somewhat mitigate the $U$-dependence of the binding affinity. This point notwithstanding, the preference to binding with NO over CO, and that over $O_2$, for moderate $U$ values, fully accords with the known behaviour of this system. The order of binding preference eventually changes with increasing $U$, but only in the unphysical region of positive interaction energies.

Binding is less favourable for each molecule when using NGWF projectors and this trend increases with increasing $U$ value. We observe approximately linear

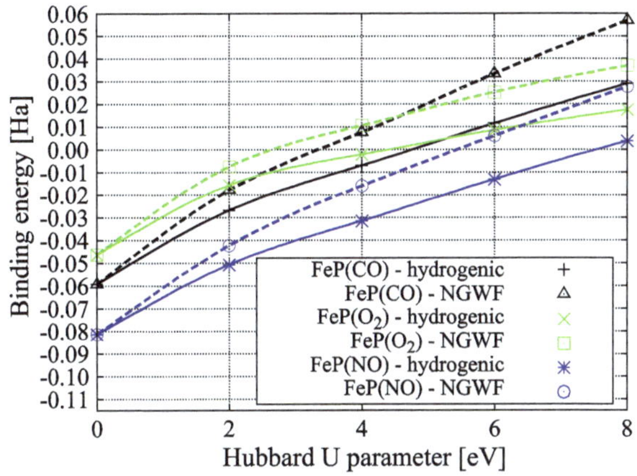


**Fig. 3.12** Binding energy of the CO, $O_2$ and NO molecules to FeP at PBE plotted as a function of $U$. All energies are calculated at PBE+$U$=0 eV geometries for the ground spin-states of FeP, FeP(CO), FeP($O_2$) and FeP(NO), and each of the respective diatomic reagents. *Solid lines* show the results calculated using hydrogenic projectors corresponding to effective charge $Z = 11.17$. *Dashed lines* show the binding energies calculated at Hubbard projector self-consistency

behaviour in the binding energy curve beyond a certain (system and projector dependent) $U$ value. This is because, beyond that value, the Kohn–Sham orbitals have effectively taken on a form very close to that of the Hubbard projectors. Thereafter, the response of the system stiffens and the energy of both reagent and product simply increases linearly with the $U$ parameter.

## 3.7 Convergence of the Projector Self-Consistency Algorithm

Figure 3.13 demonstrates the stable convergence of the projector self-consistent DFT+$U$ method for FeP+CO for a range of $U$ parameters. The graph shows the difference between the total energy at a given projector self-consistency step, at various values of $U$, and that at which the convergence criterion was satisfied—that is an energy change of less than $10^{-8}$ Ha maintained over three iterations. Each data point represents an individual variational total-energy minimisation, wherein the Hubbard projectors are re-constructed from the optimised ground-state NGWFs from the previous iteration.

The energy decreases rapidly and stably as the Hubbard projectors are refined, a consequence of the NGWFs themselves being optimised by an energy minimisation calculation at each projector iteration. This confirms our understanding that the NGWFs are well-adapted for the hybridised character of the Kohn–Sham orbitals, providing a representation of the density which reduces the DFT+$U$ energy term.

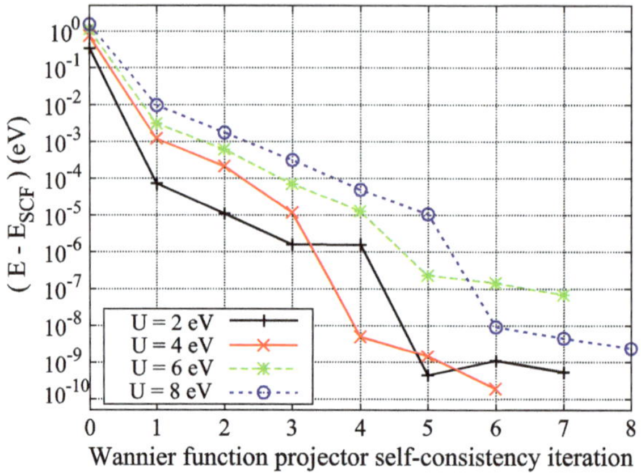


**Fig. 3.13** The difference in total energy $E$ and the total energy at projector self-consistency $E_{SCF}$ as a function of the projector self-consistency iteration. The procedure is initialised (iteration 0) with a set of hydrogenic $3d$ projectors to construct the correlated subspace, using the Clementi–Raimondi [20] effective charge of $Z = 11.17$ for iron $3d$ orbitals

Since the NGWF representation is then also optimal for the densities associated with the correlated subspaces, at projector self consistency, it provides a good description of the subspace occupancies. In this way, more spatially diffuse self-interaction corrections are introduced than with purely hydrogenic Hubbard projectors, in a complementary manner to such methods as DFT+$U$+$V$ [43] which allow more general interaction terms between sites.

The change to the total energy when the set of projectors is updated is naturally greater with a greater $U$ parameter, so the total-energy convergence of the projector self-consistency algorithm is somewhat slower for larger values of $U$. While the total-energy is not guaranteed to decrease at each successive projector update iteration, in practice, we find that it usually decreases monotonically, and approximately exponentially, as the projectors are updated. We have found that, unlike many other non-linear systems where previous histories of solutions are mixed in to damp oscillatory behaviour, the projector update scheme is very robust in this sense and no mixing with previous Hubbard projectors is required.

## 3.8 Computational Cost of Projector Self-Consistency

The ground-state density from one Hubbard projector iteration is used to initialise that of the following iteration and so much fewer NGWF optimisation steps are required at each successive projector update step. As depicted in Fig. 3.14, this results in an overall computational effort for achieving projector self-consistency that is only a small overhead compared to the conventional DFT+$U$ approach.

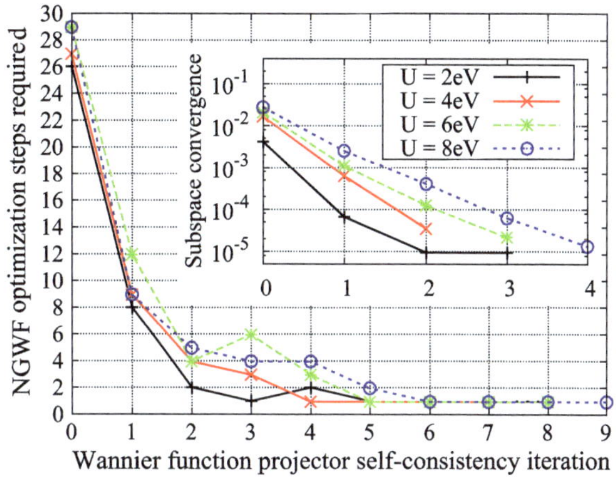


**Fig. 3.14** The number of NGWF optimisation steps needed to converge the total energy for each projector self-consistency iteration for FeP+CO. Shown inset is the convergence of the correlated subspace, as quantified by its $3d$-orbital character

Inset in Fig. 3.14, the trace of the product between the NGWF Hubbard projection at each self-consistency step and the hydrogenic initial guess projection, is plotted as the difference from its stationary value. The stationary point is defined as that at which the trace of the NGWF-hydrogenic projector product varies by less than $10^{-8}$ electrons. This shows that the correlated subspace actually converges in very few iterations indeed, with close to ideal exponential behaviour. The subsequent decreases in the total energy shown in the principal graph are thus primarily due to further refinement of the density, with effectively converged Hubbard projectors.

## 3.9 Forces in Projector Self-Consistent DFT+$U$

In order to achieve meaningful insight into the $U$-dependence of bond formation, it is necessary to allow for Hubbard projector update consistent with variations in the molecular geometry. As such, if one were to re-optimise the Hubbard projectors at each step of a molecular dynamics simulation, however, one expects that additional Pulay terms in the ionic forces might be required in order to account for changes in the spatial profile of the Hubbard projectors with ionic displacement. It would appear a nontrivial task to predict the requisite first-order change of the form Wannier functions with atomic position.

In the particular case of the ONETEP method, where the ground-state at any given atomic configuration the energy in is, by definition, variationally minimised with respect to the expansion coefficients of the NGWFs, the force calculation is simplified if a subset of these converged NGWFs are those selected as Hubbard projectors. We may assume that the forces are computed only when the energy is strictly stationary with respect to the NGWFs, as they should be, and that the NGWFs which are selected as Hubbard projectors do not change from one ionic configuration to the next; this is most likely and may be easily confirmed.

Since the Hubbard projectors are themselves NGWFs which minimise the energy, the first derivative of the energy with respect to the expansion coefficients of the Hubbard projectors vanishes also, in a similar manner to the vanishing change in the Kohn–Sham orbitals to first-order in ionic position. As a result, no first-order terms appear in the ionic forces in addition to the usual derivative of the projectors with respect to ionic position which appears in conventional DFT+$U$, given by Eqs. 2.65 and 4.32.

## 3.10 Concluding Remarks

In conclusion, we have proposed and demonstrated a method within DFT+$U$ for obtaining Hubbard projectors that are unambiguously-defined, optimally adapted to their chemical environment, and consistent with the DFT+$U$ ground-state density. Our implementation may be incorporated into any method that either solves directly for localised Wannier-like states, or which computes such states in a post-processing fashion. If combined self-consistently with approaches for calculating $U$ from first-principles [6, 7, 13], which we embellish further in Chap. 7, this work opens up the possibility of parameter-free DFT+$U$ calculations for large systems.

## References

1. D.D. O'Regan, N.D.M. Hine, M.C. Payne, A.A. Mostofi, Projector self-consistent DFT+U using nonorthogonal generalized Wannier functions. Phys. Rev. B **82**(8), 081102 (2010)
2. V.I. Anisimov, J. Zaanen, O.K. Andersen, Band theory and Mott insulators: Hubbard U instead of Stoner I. Phys. Rev. B **44**(3), 943 (1991)
3. V.I. Anisimov, I.V. Solovyev, M.A. Korotin, M.T. Czyżyk, G.A. Sawatzky, Density-functional theory and NiO photoemission spectra. Phys. Rev. B **48**(23), 16929 (1993)
4. V.I. Anisimov, A.I. Poteryaev, M.A. Korotin, A.O. Anokhin, G. Kotliar, First-principles calculations of the electronic structure and spectra of strongly correlated systems: dynamical mean-field theory. J. Phys. Condens. Matter **9**(35), 7359 (1997)
5. A.I. Lichtenstein, M.I. Katsnelson, Ab initio calculations of quasiparticle band structure in correlated systems: LDA++ approach. Phys. Rev. B **57**(12), 6884 (1998)
6. M. Cococcioni, S. de Gironcoli, Linear response approach to the calculation of the effective interaction parameters in the LDA + U method. Phys. Rev. B **71**(3), 035105 (2005)
7. H.J. Kulik, M. Cococcioni, D.A. Scherlis, N. Marzari, Density functional theory in transition-metal chemistry: a self-consistent Hubbard U approach. Phys. Rev. Lett. **97**(10), 103001 (2006)
8. T. Miyake, F. Aryasetiawan, Screened Coulomb interaction in the maximally localized Wannier basis. Phys. Rev. B **77**(8), 085122 (2008)
9. S.L. Dudarev, G.A. Botton, S.Y. Savrasov, C.J. Humphreys, A.P. Sutton, Electron-energy-loss spectra and the structural stability of nickel oxide: an LSDA+U study. Phys. Rev. B **57**(3), 1505 (1998)
10. O. Eriksson, J.M. Wills, M. Colarieti-Tosti, S. Lebgue, A. Grechnev, Many-body projector orbitals for electronic structure theory of strongly correlated electrons. Int. J. Quantum Chem. **105**, 160–165 (2005)

11. D.D. O'Regan, M.C. Payne, A.A. Mostofi, Subspace representations in ab initio methods for strongly correlated systems. Phys. Rev. B **83**(24), 245124 (2011)
12. A.A. Mostofi, J.R. Yates, Y.-S. Lee, I. Souza, D. Vanderbilt, N. Marzari, Wannier90: a tool for obtaining maximally localised Wannier functions. Comp. Phys. Comm. **178**, 685 (2008)
13. W.E. Pickett, S.C. Erwin, E.C. Ethridge, Reformulation of the LDA+U method for a local-orbital basis. Phys. Rev. B **58**(3), 1201 (1998)
14. M.J. Han, T. Ozaki, J. Yu, O(N) LDA + U electronic structure calculation method based on the nonorthogonal pseudoatomic orbital basis. Phys. Rev. B **73**(4), 045110 (2006)
15. K.K.H. Eschrig, I. Chaplygin, Density functional application to strongly correlated electron systems. J. Solid State Chem. **176**(2), 482 (2003)
16. S. Fabris, S. de Gironcoli, S. Baroni, G. Vicario, G. Balducci, Reply to comment on taming multiple valency with density functionals: a case study of defective ceria. Phys. Rev. B **72**(23), 237102 (2005)
17. R.D. King-Smith, M.C. Payne, J.S. Lin, Real-space implementation of nonlocal pseudopotentials for first-principles total-energy calculations. Phys. Rev. B **44**(23), 13063 (1991)
18. S.J. Clark, M.D. Segall, C.J. Pickard, P.J. Hasnip, M.J. Probert, K. Refson, M.C. Payne, First principles methods using CASTEP. Zeitschrift für Kristallographie **220**(5–6), 567 (2005)
19. P. Giannozzi, S. Baroni, N. Bonini, M. Calandra, R. Car, C. Cavazzoni, D. Ceresoli, G.L. Chiarotti, M. Cococcioni, I. Dabo, A.D. Corso, S. de Gironcoli, S. Fabris, G. Fratesi, R. Gebauer, U. Gerstmann, C. Gougoussis, A. Kokalj, M. Lazzeri, L. Martin-Samos, N. Marzari, F. Mauri, R. Mazzarello, S. Paolini, A. Pasquarello, L. Paulatto, C. Sbraccia, S. Scandolo, G. Sclauzero, A.P. Seitsonen, A. Smogunov, P. Umari, R.M. Wentzcovitch, Quantum espresso: a modular and open-source software project for quantum simulations of materials. J. Phys. Condens. Matter **21**(39), 395502 (2009)
20. E. Clementi, D.L. Raimondi, Atomic screening constants from SCF functions. J. Chem. Phys. **38**, 2686 (1963)
21. N. Marzari, D. Vanderbilt, Maximally localized generalized Wannier functions for composite energy bands. Phys. Rev. B **56**(20), 12847 (1997)
22. I. Souza, N. Marzari, D. Vanderbilt, Maximally localized Wannier functions for entangled energy bands. Phys. Rev. B **65**(3), 035109 (2001)
23. F. Lechermann, A. Georges, A. Poteryaev, S. Biermann, M. Posternak, A. Yamasaki, O.K. Andersen, Dynamical mean-field theory using Wannier functions: a flexible route to electronic structure calculations of strongly correlated materials. Phys. Rev. B **74**(12), 125120 (2006)
24. V.V. Mazurenko, S.L. Skornyakov, A.V. Kozhevnikov, F. Mila, V.I. Anisimov, Wannier functions and exchange integrals: the example of $LiCu_2O_2$. Phys. Rev. B **75**, 224408 (2007)
25. M. Posternak, A. Baldereschi, S. Massidda, N. Marzari, Maximally localised Wannier functions in antiferromagnetic MnO within the FLAPW formalism. Phys. Rev. B **65**, 184422 (2002)
26. B. Amadon, F. Lechermann, A. Georges, F. Jollet, T.O. Wehling, A.I. Lichtenstein, Plane-wave based electronic structure calculations for correlated materials using dynamical mean-field theory and projected local orbitals. Phys Rev B **77**(20), 205112 (2008)
27. C.-K. Skylaris, P.D. Haynes, A.A. Mostofi, M.C. Payne, Introducing ONETEP: linear-scaling density functional simulations on parallel computers. J. Chem. Phys. **122**, 084119 (2005)
28. P.D. Haynes, C.-K. Skylaris, A.A. Mostofi, M.C. Payne, Elimination of basis set superposition error in linear-scaling density-functional calculations with local orbitals optimised in situ. Chem. Phys. Lett. **422**, 345 (2006)
29. R. McWeeny, Some recent advances in density matrix theory. Rev. Mod. Phys. **32**(2), 335 (1960)
30. X.-P. Li, R.W. Nunes, D. Vanderbilt, Density-matrix electronic-structure method with linear system-size scaling. Phys. Rev. B **47**(16), 10891 (1993)
31. R.W. Nunes, D. Vanderbilt, Generalization of the density-matrix method to a nonorthogonal basis. Phys. Rev. B **50**(23), 17611 (1994)
32. M.S. Daw, Model for energetics of solids based on the density matrix. Phys. Rev. B **47**(16), 10895 (1993)

33. D.G. Anderson, Iterative procedures for nonlinear integral equations. J. ACM **12**(4), 547 (1965)
34. C.G. Broyden, A class of methods for solving nonlinear simultaneous equations. Math. Comput. **19**(92), 577 (1965)
35. D.A. Scherlis, M. Cococcioni, P. Sit, N. Marzari, Simulation of heme using DFT+U: a step toward accurate spin-state energetics. J. Phys. Chem. B **111**(25), 7384 (2007)
36. M. Bernien, J. Miguel, C. Weis, M.E. Ali, J. Kurde, B. Krumme, P.M. Panchmatia, B. Sanyal, M. Piantek, P. Srivastava, K. Baberschke, P.M. Oppeneer, O. Eriksson, W. Kuch, H. Wende, Tailoring the nature of magnetic coupling of Fe-porphyrin molecules to ferromagnetic substrates. Phys. Rev. Lett. **102**(4), 047202 (2009)
37. P.M. Panchmatia, B. Sanyal, P.M. Oppeneer, GGA+U modeling of structural, electronic, and magnetic properties of iron porphyrin-type molecules. Chem. Phys. **343**(1), 47 (2008)
38. A.M.P. Sena, V. Brázdová, D.R. Bowler, Density functional theory study of the iron-based porphyrin haem(b) on the Si(111):H surface. Phys. Rev. B **79**(24), 245404 (2009)
39. L.G.G.V. Dias da Silva, M.L. Tiago, S.E. Ulloa, F.A. Reboredo, E. Dagotto, Many-body electronic structure and Kondo properties of cobalt-porphyrin molecules. Phys. Rev. B **80**(15), 155443 (2009)
40. M. Palummo, C. Hogan, F. Sottile, P. Bagalá, A. Rubio, A binitio electronic and optical spectra of free-base porphyrins: the role of electronic correlation. J. Chem. Phys. **131**(8), 084102 (2009)
41. J.P. Perdew, K. Burke, M. Ernzerhof, Generalized gradient approximation made simple. Phys. Rev. Lett. **77**(8), 3865 (1996)
42. A.M. Rappe, K.M. Rabe, E. Kaxiras, J.D. Joannopoulos, Optimized pseudopotentials. Phys. Rev. B **41**(2), 1227 (1990)
43. V.L. Campo Jr, M. Cococcioni, Extended DFT+U+V method with on-site and inter-site electronic interactions. J. Phys. Condens. Matter **22**(5), 055602 (2010)

# Chapter 4
# Subspace Representations in Ab Initio Methods for Strongly Correlated Systems

We present, in this chapter, a generalised definition of subspace occupancy matrices in ab initio methods for strongly correlated materials, such as DFT + $U$ and DFT+DMFT, which is appropriate to the case of nonorthogonal projector functions.

By enforcing the tensorial consistency of all matrix operations, we are led to a subspace projection operator for which the occupancy matrix is tensorial and accumulates only contributions which are local to the correlated subspace at hand. For DFT + $U$ in particular, the resulting contributions to the potential and ionic forces are automatically Hermitian, without resort to symmetrisation, and localised to their corresponding correlated subspace. The tensorial invariance of the occupancies, energies and ionic forces is preserved. We illustrate the effect of this formalism in a projector-self consistent DFT + $U$ study of a representative pair of strongly correlated systems. We refer the reader to Chap. 3 and Ref. [1] for a description of the projector-self consistent DFT + $U$ method. This chapter has been published in Ref. [2]. Reprinted with permission from David D. O'Regan, Mike C. Payne and Arash A. Mostofi, Phys. Rev. B 83, 245124 (2011). Copyright (2011) by the American Physical Society.

## 4.1 Motivation

The definition of the correlated subspace occupancy matrices for the DFT + Hubbard $U$ (DFT + $U$) [3, 4] or DFT+dynamical mean field theory (DFT+DMFT) [5, 6] methods is quite unambiguous when using a set of orthonormal projectors and it is described in Chap. 2. The question of how to properly extend the formalism to allow for the possibility of nonorthogonal spanning functions, however, is one under active debate [7, 8] and one of immediate practical consequence.

It is frequently useful to permit the nonorthogonality of the basis functions for the Kohn–Sham [9] states in ab initio methods which make use of sophisticated spatially-localised orbitals for such functions, particularly in linear-scaling density functional theory methods [7, 10–12]. Additionally, either for reasons of computational convenience, as in Refs. [7, 13, 14], or for the purposes of achieving self-consistency over

D. D. O'Regan, *Optimised Projections for the Ab Initio Simulation of Large and Strongly Correlated Systems*, Springer Theses, DOI: 10.1007/978-3-642-23238-1_4, 

the correlated subspaces, as in Ref. [1] and Chap. 3, it is common to use a subset of these nonorthogonal basis functions as projectors for the correlated subspaces, the subset termed *Hubbard projectors*. As we demonstrate below, however, it may be hazardous to over-identify the Hubbard projectors with the basis set from which they are drawn.

In this chapter, we offer a revised definition of the subspace occupancy matrix for ab initio methods which use nonorthogonal projectors to define the strongly correlated subspaces. We show that by enforcing the tensorial consistency (see Chap. 5 for further detail on the significance of this) of all matrix operations we are led immediately to a simple definition of the projection operator, for each subspace, which is fully localised to that subspace. In contrast to previously proposed formalisms of Ref. [7] and references therein, this gives rise to Hermitian corrections to the potential and ionic forces, without any post hoc symmetrisation, which are also localised to the spaces in which the correlation correction is required. The resulting occupancy matrix reproduces the electron number of the subspaces and is tensorial. Thus, for example, its trace is invariant under both unitary rotations and the generalised Löwdin transformations [15] described in Ref. [8].

In order to illustrate the performance of the proposed formalism, we applied it to the DFT + $U$ method in a study of two strongly correlated systems, namely bulk nickel oxide and the gas-phase copper phthalocyanine dimer, and compared our predictions to those resulting from the most comprehensive alternative formalism available at the time of writing, the "dual representation" of Ref. [7].

A set of nonorthogonal generalised Wannier functions [10] optimised using the projector self-consistent DFT + $U$ method, described in Ref. [1] and Chap. 3, was used in order to carry out our computational study with a minimum of user intervention in the construction of the nonorthogonal Hubbard projectors.

## 4.2 Nonorthogonal Representations of the Occupancy Matrix

Generally, in order to extract low-energy Hubbard-model like models from ab initio DFT simulations, as in the DFT + $U$ method described in Chap. 2, we require the projection of the single-particle density-matrix

$$\hat{\rho}^{(\sigma)} = \sum_{i\mathbf{k}} |\psi_{i\mathbf{k}}^{(\sigma)}\rangle f_{i\mathbf{k}}^{(\sigma)} \langle \psi_{i\mathbf{k}}^{(\sigma)}|, \tag{4.1}$$

where $\psi_{i\mathbf{k}}^{(\sigma)}$ is a Kohn–Sham eigenstate for spin channel $\sigma$ with band index $i$, crystal momentum $\mathbf{k}$ and occupancy $f_{i\mathbf{k}}^{(\sigma)}$, onto a set of spatially localised subspaces. These subspaces $\mathcal{C}^{(I)}$, where $I$ is the site index, encompass that part of the Hilbert space of the Kohn–Sham orbitals which is deemed to be responsible for strong localised Coulomb interactions beyond the scope of the approximate exchange-correlation functional.

The occupancy of subspace $\mathcal{C}^{(I)}$, which is delineated by a set of $M^{(I)}$ potentially *non-orthogonal* spanning projectors $|\varphi_m^{(I)}\rangle$, $m \in \{1, \ldots, M^{(I)}\}$, dubbed *Hubbard projectors*, which are associated with subspace $I$, is generally given by the subspace-projected density matrix

$$\hat{n}^{(I)(\sigma)} = \hat{P}^{(I)\dagger} \hat{\rho}^{(\sigma)} \hat{P}^{(I)}. \tag{4.2}$$

The Hubbard projection operator $\hat{P}^{(I)}$, the resolution of the identity for the space $\mathcal{C}^{(I)}$, is defined in terms of the Hubbard projectors, but the exact manner in which this definition should be made has been the subject of some discussion, as we describe in the following.

Some important conditions should be satisfied by a sound definition of the occupancy matrix of each correlated site, namely: all operations such as matrix products and traces should be tensorially consistent so that the total energy, potential and forces are tensorial invariants (unaltered by arbitrary transformations of the basis on which the projectors for that site are defined); any potential depending on that occupancy matrix should be Hermitian and its action should be strictly localised to the correlated subspace while depending only on occupancies which are themselves localised to that subspace; the trace of the occupancy matrix should exactly reproduce the occupancy of the correlated manifold on that site and if the site is extended to encompass the entire system then the total electron number should be obtained.

### 4.2.1 The "Full" and "On-Site" Representations

We generally assume that a set of complex, mutually nonorthogonal Hubbard projectors are used for each individual site and that the correlated subspaces possibly overlap (we do not consider transformations among the projectors of different correlated sites). Dual vectors of the Hubbard projectors must be defined with respect to some Hilbert superspace of the correlated manifold, $\mathcal{H}^{(I)} \supseteq \mathcal{C}^{(I)}$, some possibilities for which are the subspace itself (i.e., $\mathcal{H}^{(I)} = \mathcal{C}^{(I)}$), the union of all correlated subspaces (i.e., $\mathcal{H}^{(I)} = \bigcup_I \mathcal{C}^{(I)}$) and the space $\mathcal{S}$ spanned by all basis functions in the simulation cell (i.e., $\mathcal{H}^{(I)} = \mathcal{S}$). The Hubbard projector duals are then generally given by

$$|\varphi^{(I)m}\rangle = \sum_{\alpha \in \mathcal{H}^{(I)}} |\varphi_\alpha^{(I)}\rangle S^{(I)\alpha m}, \tag{4.3}$$

where $S^{(I)\bullet\bullet}$ is the contravariant metric tensor for the set of functions spanning $\mathcal{H}^{(I)}$ (the inverse of their overlap matrix). Physically meaningful inner products, e.g., tensorial invariants such as occupancies, energies or forces, are computed between functions and elements of their set of dual functions only (in the orthonormal case there is no practical distinction between functions and their duals). For a more detailed exposition of tensor calculus applied to problems in electronic structure theory we refer the reader to Chap. 5 and Refs. [16, 17].

In this general case it is immediately clear that the simplest definition of the occupancy matrix for a given site, that is the projection $\hat{P}^{(I)} = \sum_{m \in \mathcal{C}^{(I)}} |\varphi_m^{(I)}\rangle\langle\varphi_m^{(I)}|$ of the valence manifold over the site's Hubbard projectors,

$$n_{mm'}^{(I)(\sigma)} = \langle\varphi_m^{(I)}|\hat{\rho}^{(\sigma)}|\varphi_{m'}^{(I)}\rangle, \tag{4.4}$$

is invalid for nonorthogonal projectors. This widely-used definition of the occupancy matrix, which is entirely appropriate in the orthonormal case such as calculations described in Ref. [18] and numerous citations therein, simply neglects all such nonorthogonality, and the trace or powers of such a fully covariant tensor are not physically meaningful in the nonorthogonal case.

A total site occupancy defined as a trace operation on this matrix, as in

$$N^{(I)(\sigma)} = \sum_m n_{mm}^{(I)(\sigma)}, \tag{4.5}$$

implies that such an occupancy is not, in general, a tensorial invariant since it is formed by a tensorially invalid summation over two covariant indices—as opposed to a meaningful contraction of indices of opposite tensor character. Occupancies, just like total energies, should be tensorial invariants, scalars which are unchanged by transformations of the basis on which the projector functions are defined.

Progress was made in the definition of the occupancies of correlated subspaces via nonorthogonal projectors when it was noted [14] that tensorially contravariant projector duals should be involved, a concept known in other contexts for some time [19]. A definition of the occupancy matrix fully in terms of Hubbard projector duals was described in Ref. [14], for example, where the projection operator defined as $\hat{P}^{(I)} = \sum_{m \in \mathcal{C}^{(I)}} |\varphi^{(I)m}\rangle\langle\varphi^{(I)m}|$, provides an occupancy matrix

$$\begin{aligned} n^{(I)(\sigma)mm'} &= \langle\varphi^{(I)m}|\hat{\rho}^{(\sigma)}|\varphi^{(I)m'}\rangle \\ &= S^{(I)m\alpha}\langle\varphi_\alpha^{(I)}|\hat{\rho}^{(\sigma)}|\varphi_\beta^{(I)}\rangle S^{(I)\beta m'}. \end{aligned} \tag{4.6}$$

The indices $\alpha$ and $\beta$ run over the spanning vectors of the contravariant metric (i.e., the inverse overlap matrix) $S^{(I)\bullet\bullet}$, on a superspace $\mathcal{H}^{(I)}$ of the correlated manifold $\mathcal{C}^{(I)}$, and we hereafter make use of the summation convention [20], whereby repeated indices within the same expression are summed over unless in parentheses.

With projector duals defined this way, the indices $\alpha$ and $\beta$ run over the spanning vectors of the contravariant metric (i.e., the inverse overlap matrix) $S^{\bullet\bullet}$ on a superspace $\mathcal{H}^{(I)}$ of the correlated manifold $\mathcal{C}^{(I)}$. Unfortunately, the resulting matrix trace and powers are again not tensorially valid, as can be seen by taking the example of the square of this contravariant occupancy matrix, which is of interest for density–density self-interaction corrections to exchange-correlation functionals. The resulting expression for the squared occupancy matrix

$$n^{2(I)(\sigma)mm'} = \sum_{m'' \in \mathcal{C}^{(I)}} n^{(I)(\sigma)mm''} n^{(I)(\sigma)m''m'} \tag{4.7}$$

implies that the operator

$$\hat{P}^{(I)} = \sum_{m'' \in \mathcal{C}^{(I)}} |\varphi^{(I)m''}\rangle\langle\varphi^{(I)m''}| \tag{4.8}$$

forms a tensorially traceable identity on $\mathcal{C}^{(I)}$. Unfortunately, it does not in the case of nonorthogonal projectors, since an identity operator can only be formed via the outer product between a projector and a projector dual, and not a dual vector and its complex conjugate, including the case where the correlated subspace is extended to the Hilbert space of the entire system ($\mathcal{C}^{(I)} = \mathcal{S}$).

The shortcomings in the two definitions of the occupancy matrix described above have been previously described in detail by Han et. al. in Ref. [7] and are dubbed, respectively, the "full" (Eq. 4.4) and "on-site" (Eq. 4.6) representations in the nomenclature described therein. These authors concentrated on the special case where the dual-generating superspace $\mathcal{H}^{(I)}$ is the space spanned by all basis functions $\{|\phi_\alpha\rangle\}$ in the simulation cell, so that $\mathcal{H}^{(I)} = \mathcal{S}$, in which case $S_{\alpha\beta} = \langle\phi_\alpha|\phi_\beta\rangle$ and the Hubbard projectors form a subset of the basis set. Thus, the same contravariant metric for all basis functions in the simulation cell is used to generate the dual functions on each correlated site and in this case the "full" and "on-site" occupancy matrices simplify, respectively, to

$$n^{(I)(\sigma)}_{mm'} = \sum_{\alpha,\beta \in \mathcal{S}} S^{(I)}_{m \in \mathcal{C}^{(I)} \alpha} K^{(\sigma)\alpha\beta} S^{(I)}_{\beta m' \in \mathcal{C}^{(I)}} \tag{4.9}$$

and

$$n^{(I)(\sigma)mm'} = K^{(\sigma)m \in \mathcal{C}^{(I)} m' \in \mathcal{C}^{(I)}} \tag{4.10}$$

where $K^{(\sigma)\alpha\beta} = \langle\phi^\alpha|\hat{\rho}^{(\sigma)}|\phi^\beta\rangle$ is the representation of the density matrix in terms of basis-set duals. The notation $S^{(I)}_{m \in \mathcal{C}^{(I)} \alpha}$ reminds us that $m$ and $\alpha$ run over the spanning vectors of two different spaces, $\mathcal{C}^{(I)}$ and $\mathcal{H}^{(I)}$, respectively, so that the block of $S_{\bullet\bullet}$ in question is generally not square.

### 4.2.2 The "Dual" Representation

Han et. al. [7], whose invaluable contribution on this subject addressed many of the salient issues, pointed out that the total number of electrons is not recovered by the trace of the occupancy matrix if the site is extended to include the entire simulation cell using the "full" and "on-site" representations, and proposed an alternative "dual" representation which solves this particular problem and is generated by the projector

$$\hat{P}^{(I)} = \frac{1}{2} \sum_{m \in \mathcal{C}^{(I)}} \left( |\varphi^{(I)m}\rangle\langle\varphi_m^{(I)}| + |\varphi_m^{(I)}\rangle\langle\varphi^{(I)m}| \right) \quad (4.11)$$

and the corresponding occupancy matrix

$$\begin{aligned} \mathbf{n}^{(I)(\sigma)} &= \frac{1}{2} \begin{pmatrix} \langle\varphi^{(I)m}|\hat{\rho}^{(\sigma)}|\varphi_{m'}^{(I)}\rangle + \\ \langle\varphi_m^{(I)}|\hat{\rho}^{(\sigma)}|\varphi^{(I)m'}\rangle \end{pmatrix} \\ &= \frac{1}{2} \sum_{\alpha \in \mathcal{S}} \begin{pmatrix} K^{(\sigma)m \in \mathcal{C}^{(I)}\alpha} S_{\alpha m' \in \mathcal{C}^{(I)}} + \\ S_{m \in \mathcal{C}^{(I)}\alpha} K^{(\sigma)\alpha m' \in \mathcal{C}^{(I)}} \end{pmatrix}. \end{aligned} \quad (4.12)$$

Here, the contravariant metric on the complete basis set is used to form the Hubbard projector duals (which are therefore delocalised across the entire simulation cell, in general, since the inverse overlap matrix is dense even when the overlap matrix itself is sparse) via $|\varphi^{(I)m}\rangle = \sum_{\alpha \in \mathcal{S}} |\varphi_\alpha^{(I)}\rangle S^{\alpha m}$. Symmetrisation is carried out in a post hoc step in order to both provide a symmetric occupancy matrix and to retrieve a Hermitian potential.

The "dual" representation shares with the "full" representation the attributes of Hermiticity and rotational invariance and, furthermore, it has a tensorially and physically meaningful trace. As such, to our knowledge, it provides the most favourable occupancy definition hitherto available. However, this occupancy matrix is tensorially ambiguous, consisting of the sum of tensors of differing index character.

One cannot generally symmetrise or antisymmetrise a tensor over indices of mixed covariant/contravariant character in this way and obtain a matrix which transforms as a tensor (and be used to generate an invariant occupancy or energy). Thus, while providing a significant improvement over previously suggested definitions of the occupancy matrix due to its tensorially invariant trace, the "dual" representation suffers similar problems with matrix powers as other representations: if we attempt to compute the square of this matrix we obtain tensorially inconsistent, and thus physically meaningless, terms in the product of the form $n^{\bullet}{}_{\bullet} n_{\bullet}{}^{\bullet}$ and $n_{\bullet}{}^{\bullet} n^{\bullet}{}_{\bullet}$.

### 4.2.3 *Requirement for a Subspace-Localised Hermitian Projection Operator*

Let us step back for a moment and consider why the projection operator

$$\hat{P}^{(I)} = \sum_{m \in \mathcal{C}^{(I)}} |\varphi^{(I)m}\rangle\langle\varphi_m^{(I)}| = \sum_{\substack{m \in \mathcal{C}^{(I)} \\ \alpha \in \mathcal{H}^{(I)} \neq \mathcal{C}^{(I)}}} |\varphi_\alpha^{(I)}\rangle S^{\alpha m} \langle\varphi_m^{(I)}| \quad (4.13)$$

requires symmetrisation to the "dual" form in order to provide a Hermitian potential operator. An arbitrary potential operator $\hat{V}$, operating on the subspace $\mathcal{C}^{(I)}$, which could represent the screened Coulomb interaction, for example, has matrix elements in the frame of Hubbard projectors given by

$$V_m^{(I)m'} = \sum_{\alpha \in \mathcal{H}^{(I)}} \langle \varphi_m^{(I)} | \hat{V} | \varphi_\alpha^{(I)} \rangle S^{\alpha m'}. \tag{4.14}$$

The potential operator is thus easily shown to be non-Hermitian in the case where $m, m' \in \mathcal{C}^{(I)} \subset \mathcal{H}^{(I)} \subseteq \mathcal{S},\ \mathcal{C}^{(I)} \neq \mathcal{H}^{(I)}$, since

$$\begin{aligned} \hat{V}^{(I)} &= \hat{P}^{(I)\dagger} \hat{V} \hat{P}^{(I)} = |\varphi_m^{(I)}\rangle V^{(I)mm'} \langle \varphi_{m'}^{(I)}| \\ &= \sum_{\alpha,\beta \in \mathcal{H}^{(I)}} |\varphi_m^{(I)}\rangle S^{(I)m\alpha} V_{\alpha\beta}^{(I)} S^{(I)\beta m'} \langle \varphi_{m'}^{(I)}| \\ &\neq \sum_{\alpha,\beta \in \mathcal{H}^{(I)}} |\varphi_\alpha^{(I)}\rangle S^{(I)\alpha m} V_{mm'}^{(I)} S^{(I)m'\beta} \langle \varphi_\beta^{(I)}| \\ &= |\varphi^{(I)m}\rangle V_{mm'}^{(I)} \langle \varphi^{(I)m'}| = \hat{P}^{(I)} \hat{V} \hat{P}^{(I)\dagger} = \hat{V}^{(I)\dagger}. \end{aligned} \tag{4.15}$$

The reason for this non-Hermiticity is that the indices $\alpha, \beta$ do not generally run over functions spanning just the correlated space $\mathcal{C}^{(I)}$, but rather over those that span a superspace $\mathcal{H}^{(I)}$, e.g., typically over the basis functions in the simulation cell, $\mathcal{H}^{(I)} = \mathcal{S}$. This observation is quite general: the dual projectors must be constructed using the metric on *precisely* the space spanned by the projectors themselves in order to build a Hermitian projection operator and hence a Hermitian potential. This cannot be circumvented in a tensorially-consistent way by symmetrising operators since tensors can only be symmetrised over pairs of indices if they are either both of covariant character or both of contravariant character.

A simple, though ultimately incomplete, work-around for this issue is to simply truncate the sum to the subspace in question when building the Hubbard projector duals, that is to define a set of approximate dual Hubbard projectors $|\varphi^{(I)m}\rangle = \sum_{m' \in \mathcal{C}^{(I)}} |\varphi_{m'}^{(I)}\rangle S^{(I)m'm}$. In effect, this corresponds to extracting, for each site $I$, a sub-block of the overlap matrix $S^{(I)\bullet\bullet}$ of size $M^{(I)} \times M^{(I)}$ (where $M^{(I)}$ is the number of Hubbard projectors for site $I$, typically 5 for a transition-metal ion) and use this matrix as an approximate contravariant metric on $\mathcal{C}^{(I)}$. The advantage of this approach is that the resulting potential is Hermitian without resorting to symmetrisation and the approximate duals are localised to the subspace to which they correspond. However, in this case the Hermitian projection

$$\hat{P}^{(I)} = \sum_{m,m' \in \mathcal{C}^{(I)}} |\varphi_{m'}^{(I)}\rangle S^{(I)m'm} \langle \varphi_m^{(I)}| \tag{4.16}$$

is, at best, an approximation to the identity operator on the subspace $\mathcal{C}^{(I)}$ and, in this case, $\langle \varphi_m^{(I)} | \varphi^{(I)m'} \rangle \neq \delta_m^{m'}$. We will denote this incomplete formulation the "truncated" representation and, while we mention it for completeness, we do not recommend its use.

### 4.2.4 The "Tensorial" Representation

Based on the above arguments, we must conclude that in order to build a tensorially-consistent occupancy matrix which generates a Hermitian potential, the projection operator for a given subspace $\mathcal{C}^{(I)}$ must be constructed using exact dual Hubbard projectors with respect to that subspace only. Thus, with the covariant overlap matrix of Hubbard projectors defined by

$$O^{(I)}_{mm'} = \langle \varphi^{(I)}_m | \varphi^{(I)}_{m'} \rangle, \tag{4.17}$$

an *individual* $M^{(I)} \times M^{(I)}$ covariant metric tensor for each correlated site $I$, the proper dual vectors $|\varphi^{(I)m}\rangle$ are constructed using the corresponding contravariant metric $O^{(I)m'm}$ (which is obtained as an individual $M^{(I)} \times M^{(I)}$ inverse operation for each site) as

$$|\varphi^{(I)m}\rangle = \sum_{m' \in \mathcal{C}^{(I)}} |\varphi^{(I)}_{m'}\rangle O^{(I)m'm}; \; O^{(I)m'm''} O^{(I)}_{m''m} = \delta^{m'}_{m}. \tag{4.18}$$

In the special case where the Hubbard projectors are drawn from the set of functions used to represent the Kohn–Sham wave-functions, the overlap matrix of duals $O^{(I)\bullet\bullet}$ for each site cannot generally be extracted immediately from the metric $S^{\bullet\bullet}$ on $\mathcal{S}$. However, in this particular case, the $O^{(I)}_{\bullet\bullet}$ matrix for each site is merely a sub-block of the basis-function overlap $S_{\bullet\bullet}$ and, from this, the contravariant $O^{(I)\bullet\bullet}$ for each site can be computed by a separate inverse operation for each site which is fast, due to the small matrix dimension.

Employing this definition of the metric tensor on each subspace, the projector duals remain manifestly as localised to the correlated subspace as the projectors themselves, they pick up only subspace-localised contributions to the occupancy and can only apply subspace-localised corrective potentials, as we would expect for local corrections such as DFT + $U$ or its extensions. The Hubbard projection operator, in what we will denote the "tensorial" representation,

$$\hat{P}^{(I)} = \sum_{m,m' \in \mathcal{C}^{(I)}} |\varphi^{(I)}_{m'}\rangle O^{m'm} \langle \varphi^{(I)}_m | \tag{4.19}$$

is Hermitian and thus gives rise to a Hermitian potential without resort to symmetrisation since $O^{(I)\bullet\bullet}$ is a square overlap matrix,

$$\begin{aligned} \hat{V}^{(I)} &= \hat{P}^{(I)\dagger} \hat{V} \hat{P}^{(I)} = |\varphi^{(I)}_m\rangle V^{(I)mm'} \langle \varphi^{(I)}_{m'}| \\ &= |\varphi^{(I)}_m\rangle O^{(I)mm'} V^{(I)}_{m'm''} O^{(I)m''m'''} \langle \varphi^{(I)}_{m'''}| \\ &= |\varphi^{(I)m}\rangle V^{(I)}_{mm'} \langle \varphi^{(I)m'}| = \hat{P}^{(I)} \hat{V} \hat{P}^{(I)\dagger} = \hat{V}^{(I)\dagger}. \end{aligned} \tag{4.20}$$

The occupancy matrix is most easily expressed in its singly covariant and singly contravariant form, though other forms are readily obtainable from the metric tensor, so manipulations of the following form can be made:

$$n_{\bullet}{}^{\bullet} = O_{\bullet\bullet} n^{\bullet\bullet} = n_{\bullet\bullet} O^{\bullet\bullet} = O_{\bullet\bullet} n^{\bullet}{}_{\bullet} O^{\bullet\bullet}. \tag{4.21}$$

The contravariant–covariant or covariant–contravariant forms of the tensorial occupancy matrix, the latter given by (the second line applies to the special case where Hubbard projectors are drawn from the basis set)

$$\begin{aligned} n_{m}^{(I)(\sigma)m'} &= \langle \varphi_{m}^{(I)} | \hat{\rho}^{(\sigma)} | \varphi_{m''}^{(I)} \rangle O^{(I)m''m'} \\ &= \sum_{\alpha\beta \in \mathcal{S}} S_{m\alpha} K^{\alpha\beta} S_{\beta m''} O^{(I)m''m'}, \end{aligned} \tag{4.22}$$

possess a common tensorially invariant trace (a proper contraction over one covariant and one contravariant index) which recovers the exact number of electrons in the correlated subspace by construction (the so-called "sum-rule"), so that

$$N^{(I)(\sigma)} = \sum_{m \in \mathcal{C}^{(I)}} n^{(I)(\sigma)m}{}_{m} = \sum_{m \in \mathcal{C}^{(I)}} n^{(I)(\sigma)}{}_{m}{}^{m}. \tag{4.23}$$

Their powers themselves remain tensors, for example the square $n^{2(I)(\sigma)m'}_{\quad m} = n^{(I)(\sigma)m''}_{\quad m} n^{(I)(\sigma)m'}_{\quad m''}$ is itself a well behaved singly covariant and singly contravariant tensor with an invariant trace. This occupancy matrix is easily demonstrated to be invariant under rotations of the set of Hubbard projectors on its site and it is independent of the basis used to represent the Kohn–Sham states.

An occupancy matrix which is invariant under element-wise transpose might lend itself to an interpretation as quantifying the charge shared between Hubbard projectors, and indeed it does in the case of orthonormal Hubbard projectors. However, it is worth emphasising that in the case of a set of nonorthogonal Hubbard projectors, these functions are merely spanning vectors with no rigorous physical meaning, and generally no such interpretation of charge shared between orbitals can be made safely. In fact, in the nonorthogonal case, the occupancy matrix should not be generally expected to be invariant under element-wise transpose, i.e., $n_{m}{}^{m'} \neq n_{m}^{T\,m'} = n^{m'}{}_{m}$. Rather, if the duals are defined in a way which preserves the tensorial consistency of inner products, the occupancy matrix must satisfy instead the more general expression $n_{m}{}^{m'} = O_{mm''} n^{m''}{}_{m'''} O^{m'''m'}$, where $O_{\bullet\bullet}$ and $O^{\bullet\bullet}$ are the covariant and contravariant metric tensors, respectively on the subspace in question. As a result, only the diagonal elements of the occupancy matrix can be imbued with a physical meaning in the sense of occupancy and symmetrising the matrix does not recover such an interpretation for the off-diagonal elements.

## 4.3 Application to the DFT + $U$ Method

In this section, we illustrate the practical application of the "tensorial" representation to a particular method for strongly correlated materials, namely the simplified rotationally invariant DFT + $U$ correction of Refs. [18, 21] that was introduced in Chap. 2.

We also bring the method for computing the $U$ parameter which is described in these articles, and which is detailed further in Chap. 7, within the tensorial framework.

It is useful, but not obligatory, to bear in mind the motivation of the DFT + $U$ correction as an idempotency penalty-functional for the subspace-projected single-particle density matrices before considering the nonorthogonal framework; for this we refer the reader to Chap. 2. Notwithstanding, in this chapter, we provide all the necessary expressions for the tensorially invariant DFT + $U$ terms in the energy, potential and ionic forces for use with nonorthogonal Hubbard projectors.

It is of some importance to demystify the use of nonorthogonal Hubbard projectors in the case of DFT + $U$ because such a set is often used in contemporary, particularly linear-scaling, implementations [7, 10–12]. We show in Chap. 3 and Ref. [1] that an efficient set of Hubbard projectors can be constructed which is self-consistent with the set of truncated nonorthogonal generalised Wannier functions that minimises the DFT + $U$ total energy.

When the necessary steps are taken to ensure the tensorial consistency of matrix operations, orthonormalisation of the Hubbard projectors before their use is rendered unnecessary. Nonorthogonality is perhaps a desirable property, in fact, for a set of Hubbard projectors in the form of Wannier functions, as it permits them to have greater localisation than their equivalent orthonormal counterparts [22].

We concentrate here on the familiar rotationally invariant simplified DFT + $U$ correction term of Cococcioni and de Gironcoli in Ref. [18], given for an orthonormal set of Hubbard projectors for each site by

$$\sum_{I\sigma} \frac{\bar{U}^{(I)(\sigma)}}{2} \left[ \sum_{m} n_{mm} - \sum_{mm'} n_{mm'} n_{m'm} \right]^{(I)(\sigma)} . \tag{4.24}$$

If we further assume that atomic projectors are employed, the scalar $\bar{U}^{(I)(\sigma)}$ represents a spherical average of that Coulomb repulsion. As described in Chap. 2, this is a simplified version of the rotationally invariant form of the DFT + $U$ functional given in reference [23]. The extension of alternative forms of the DFT + $U$ functional to our nonorthogonal formalism is straightforward.

### 4.3.1 The Tensorially Invariant DFT + U Functional

Let us consider how we might generalise this DFT + $U$ penalty functional to include orbital-dependent interactions in a rotationally invariant and tensorially consistent manner. The Coulomb interaction tensor $U$ for a given spin channel and site (considering the same Hubbard projectors for different spins for brevity of notation) is given generally by the two-centre integral (N.B. using the Dirac, and not Mulliken, convention)

$$U^{(I)}_{mm'm''m'''} = \langle \varphi^{(I)}_{m'} \varphi^{(I)}_{m'} | \hat{U}^{(I)(\sigma)} \left( \mathbf{r}, \mathbf{r}' \right) | \varphi^{(I)}_{m''} \varphi^{(I)}_{m'''} \rangle . \tag{4.25}$$

Here $\hat{U}^{(\sigma)}\left(\mathbf{r},\mathbf{r}'\right)$ is the Coulomb interaction screened according to mechanisms described by an appropriate theory such as linear response [13, 18, 21], constrained LDA [24–26], constrained RPA [27, 28] or constrained adiabatic LDA [29]. Coulomb repulsion is represented by those terms for which $m = m''$; $m' = m'''$, while direct exchange is given by those elements with $m = m'''$; $m' = m''$. We refer the reader Chap. 7 for an elaboration on this topic.

In the general, nonorthogonal case, care must be taken in computing and using the $U$ tensor in order to preserve the tensorial invariance of the DFT + $U$ energy. For example, if a tensorial invariant is required which provides the sum of the part of the tensor describing density-density Coulomb repulsions, it should correctly be computed by contracting covariant and contravariant indices in pairs of indices of opposite character, e.g., double-sums of the form, where $m, m' \in \{1, \ldots, M^{(I)}\}$,

$$U_{mm'}{}^{mm'}, \quad U^{mm'}{}_{mm'}, \quad U_{m}{}^{m'm}{}_{m'}, \quad or \quad U^{m}{}_{m'm}{}^{m'} \tag{4.26}$$

are admissible while those of the form $U_{mm'mm'}$ or $U^{mm'mm'}$ break tensorial invariance. Indices are raised and lowered using the metric tensor of the correlated subspace to which the $U$ tensor corresponds, the contravariant $O^{\bullet\bullet}$ or covariant $O_{\bullet\bullet}$, respectively, as in

$$U_{m}{}^{m'}{}_{m'}{}^{m} = O^{m'm''} U_{mm''m'm'''} O^{m'''m}. \tag{4.27}$$

A fully projector-decomposed tensorially invariant DFT + $U$ functional may be constructed using pairwise contractions over the four indices, for example

$$\sum_{I\sigma} \frac{1}{2} U^{(I)}{}_{mm''}{}^{m'm'''} \left[ n_{m'}{}^{m} \delta^{m''}_{m'''} - n_{m'}{}^{m''} n_{m'''}{}^{m} \right]^{(I)(\sigma)}, \tag{4.28}$$

which includes all information on the spurious curvature of the total energy with respect to orbital occupation up to the resolution of the Hubbard projectors.

A commonly used approximation for the screened Coulomb interaction, at the time of writing, is where the both the perturbing and probing indices are contracted over before use, providing a scalar density-density Coulomb interaction. The usual DFT + $U$ penalty-functional in this fully averaged approximation is thus given in tensorially-invariant form by the expression, where the interaction tensor is properly renormalised,

$$\sum_{I\sigma} \frac{1}{2} U^{(I)}{}_{mm'}{}^{mm'} \left[ n_{m''}{}^{m''} - n_{m''}{}^{m'''} n_{m'''}{}^{m''} \right]^{(I)(\sigma)}. \tag{4.29}$$

### 4.3.2 DFT + U Potential and Ionic Forces

The DFT + $U$ term in the Kohn–Sham potential, generally given (for real valued $U$ tensors) by

$$\hat{V}^{(\sigma)} = \sum_I |\varphi^{(I)m}\rangle V_m^{(I)(\sigma)m'} \langle \varphi_{m'}^{(I)}|, \tag{4.30}$$

has matrix elements

$$V_m^{(I)(\sigma)m'} = \frac{1}{2} U_{m''m'''}^{(I)\ m''m'''} \left[ \delta_{m'}^{m} - 2n_{m'}^{(I)(\sigma)m} \right]. \tag{4.31}$$

in the fully-averaged scalar interaction case. The DFT + $U$ potential is Hermitian by construction when the Hubbard projection operator built with the subspace-local tensorial representation of Eq. 4.19, is used. No symmetrisation of the occupancy matrices is needed to ensure this Hermiticity and the potential acts strictly within the spatial extent of the subspace of whose occupancy it depends.

Correspondingly, the DFT + $U$ contribution to the force on the ion labelled $J$, with position $\mathbf{R}_J$, is given by

$$\begin{aligned} \mathbf{F}_J = &- \sum_{I\sigma} \left\langle \frac{d\varphi_m^{(I)}}{d\mathbf{R}_J} \middle| \varphi_{m'}^{(I)} \right\rangle O^{(I)m'm''} n_{m''}^{(I)(\sigma)m'''} V_{m'''}^{(I)(\sigma)m} \\ &- \sum_{I\sigma} n_m^{(I)(\sigma)m'} \left\langle \varphi_{m'}^{(I)} \middle| \frac{d\varphi_{m''}^{(I)}}{d\mathbf{R}_J} \right\rangle O^{(I)m''m'''} V_{m'''}^{(I)(\sigma)m}, \end{aligned} \tag{4.32}$$

where, for example, we have made simplifications such as

$$\left| \frac{d\varphi^{(I)m}}{d\mathbf{R}_J} \right\rangle = \left| \frac{d\varphi_{m'}^{(J)}}{d\mathbf{R}_J} \right\rangle O^{(I)m'm}. \tag{4.33}$$

Here we have expressed the force term in the frame of Hubbard projectors and we refer the reader to Eq. 2.65 for its expression, in the case of real-valued Hubbard projectors, in terms of localised nonorthogonal real-valued support functions.

Condition Eq. 4.33 holds if the subspace metric tensor is position independent, in particular if the Hubbard projectors are simply spatially translated when their host ion is moved, or when the energy is minimised with respect to the Hubbard projectors before the force is calculated. The force equation, of course, holds only if we are on the Hellmann–Feynman surface, where the density matrix commutes with the Hamiltonian and so all terms contributing to the force involving derivatives of the density-matrix with respect to ionic positions vanish.

### 4.3.3 The Case of Orthonormal Hubbard Projectors

Orthogonal sets of Hubbard projectors, as well as nonorthogonal sets, may provide a compact and accurate representation of the correlated subspaces and we would not wish to detract from their value and ease of use. In this special case, the Hubbard projectors equal their own duals with respect to their subspace, and the metric tensor reduces to a Kronecker delta. The Maximally-Localised Wannier functions [30, 31], for example, formed by minimising the quadratic spread of the Wannier functions for a set of bands in a specified window of the Kohn–Sham eigenvalue spectrum, have been shown to provide an excellent minimal basis with which to build tight-binding models using Kohn–Sham orbitals. These functions also seem to be those which maximise the four-index trace of the Coulomb interaction [32]. The $U$ tensor, or those elements relevant for screened Coulomb repulsion $U^{(I)(\sigma)}_{mm'mm'}$, is a complicated object even in the orthonormal case and one whose off-diagonal elements should not be ignored if possible.

The DFT + $U$ energy functional, in the case where one orbital degree of freedom is retained and an orthonormal set of Hubbard projectors is used for the site in question, is given by

$$\sum_{I\sigma} \frac{1}{2} U^{(I)}_{mm''m'm''} \left[ n^{(I)(\sigma)}_{mm'} - n^{(I)(\sigma)}_{m'm'''} n^{(I)(\sigma)}_{m'''m} \right]. \tag{4.34}$$

This further reduces to the commonly used form of Cococcioni and de Gironcoli [18] if we take a second trace over the $U$ tensor and disallow orbital off-diagonal interactions in the above, replacing

$$U^{(I)}_{mm''m'm''} \quad \text{with} \quad U^{(I)}_{m'''m''m'''m''} \delta_{mm'} \tag{4.35}$$

and identifying $U^{(I)} = U^{(I)}_{m'''m''m'''m''}$ as the usual subspace-averaged scalar $U$.

If one performs an inverse Löwdin transform [15] from an orthonormal set of projectors to a nonorthogonal frame using the matrix square root of covariant and contravariant metric on a particular correlated subspace, $O^{\frac{1}{2}}_{\bullet\bullet}$ and $O^{-\frac{1}{2}\bullet\bullet}$ for that $\mathcal{C}^{(I)}$, respectively, then the pre-multiplicative scalar $U$ parameter for that site remains identically the same, since for each site (if $n$ and $n'$ index orthonormal projectors and $m$ and $m'$ index their nonorthogonal counterparts) we have

$$\begin{aligned} &\sum_{nn'} U_{nn'nn'} \\ &= \sum_{nn'} U_{nn'n''n'''} \delta_{nn''} \delta_{n'n'''} \\ &= \sum_{nn'} U_{nn'n''n'''} \sum_{m} O^{\frac{1}{2}}_{nm} O^{-\frac{1}{2}mn''} \sum_{m} O^{\frac{1}{2}}_{n'm'} O^{-\frac{1}{2}m'n'''} \\ &= \sum_{mm'} U_{mm'}{}^{mm'} = U. \end{aligned} \tag{4.36}$$

Thus, in the widely used assumption in which the Coulomb interaction is replaced by a pre-multiplicative scalar times the identity, we retain the usual interpretation of the scalar $U$ as the averaged screened Coulomb repulsion between densities in the subspace described by the Hubbard projectors, or as a penalty-functional pre-factor which can be varied or computed in order to minimise the self-interaction error, regardless of whether or not the Hubbard projectors are orthonormal.

### *4.3.4 Invariance Under Generalised Löwdin Transforms*

As suggested in Ref. [8], generalised definitions of the Löwdin transform may be used where the metric tensor is raised to an arbitrary power $A$, as is its inverse, and the canonical Löwdin transform $A = \frac{1}{2}$ has the status of a special case. Since, however, by construction

$$\sum_{m \in \mathcal{C}^{(I)}} O_{nm}^{(I)(A)} O^{(I)(-A)mn''} = \delta_{nn''} = \sum_{\gamma \in \mathcal{S}} S_{n\gamma}^{(A)} S^{(-A)\gamma n''}, \tag{4.37}$$

the fully averaged scalar $U$ is invariant under such transformations, and independent of the exponent $A$, regardless of whether the subspace metric tensor $O_{\bullet\bullet}$ or, in the "dual" representation case, the metric $S_{\bullet\bullet}$ on the space of basis functions is used.

Such transforms were explored in Ref. [8] within the context of DFT + $U$ with nonorthogonal projectors and with the basis set metric tensor $S$ generating the duals (in which case projector and occupancy matrix symmetrisation was required to produce a Hermitian potential in generalisations of the "dual" representation to non-zero or non-unity values of $A$, and thus tensorial invariance was lost for the reasons we have explained).

In a sense, this generalised Löwdin transformation exponent $A$ varies the nonorthogonality of the representation of the occupancy matrices or, equivalently, (since the basis set metric $S$ introduces spurious contributions to the occupancy matrix from across the simulation cell) the degree of non-locality of the correction. The dependence of computed ground-state properties and of the Kohn–Sham gap of a variety of materials on $A$, as reported in Ref. [8], demonstrates, in our view, the ambiguity of population analysis measures, and hence corrections such as DFT + $U$, which are built from tensorially inconsistent occupancy matrices where the delocalised Hubbard projector duals of the form $|\tilde{\varphi}^{(I)m}\rangle = \sum_{\gamma \in \mathcal{S}} |\tilde{\varphi}_{\gamma}^{(I)}\rangle S^{(-A)\gamma m}$ are used in the construction of the Hubbard projector.

The conclusion that the $A$ parameter bears influence on computed properties accords well with our arguments on the unsuitability of the metric $S$ (on the basis functions in the entire simulation cell) in constructing localised self-interaction corrections such as DFT + $U$, since that parameter effectively controls the degree of spatial delocalisation of the Hubbard projector duals and hence, in a sense, the severity of the tensorial inconsistency in the DFT + $U$ functional. The occupancy matrix for the "dual" representation subject to a generalised Löwdin transformation, given by

$$\sum_{\gamma,\delta \in \mathcal{S}} S^{(A)m\gamma} \langle \varphi_{\gamma}^{(I)} | \hat{\rho} | \varphi_{\delta}^{(I)} \rangle S^{(1-A)\delta m'}, \tag{4.38}$$

picks up differing non-local contributions (densities from outside the correlated subspace) with varying $A$. Spurious non-local contributions are incorporated for all values of $A$.

On the contrary to this, however, when the metric tensor $O$ appropriate to the subspace at hand is used, the generalised Löwdin transformed occupancy matrix

$$\sum_{m'',m'''\in\mathcal{C}^{(I)}} O^{(I)(A)mm''} \langle\varphi^{(I)}_{m''}|\hat{\rho}|\varphi^{(I)}_{m'''}\rangle O^{(I)(1-A)m'''m'}, \qquad (4.39)$$

contains no contributions from outside the correlated subspace it is describing, for any value of $A$. The trace of this matrix is fully independent of $A$, since $O^{(I)(1-A)m'''m} O^{(I)(A)mm''} = O^{(I)m'''m''}$. This is the case also for the trace of the square of this matrix, so that the DFT + $U$ correction is entirely independent of $A$ by construction, and so is unambiguously defined when the appropriate subspace-local tensorial metric tensor is used to build the projection operator.

## 4.4 Strongly-Correlated Insulator: Bulk Nickel Oxide

The DFT + $U$ method has previously been applied, in numerous incarnations, to bulk NiO and it is known to recover the principal features of this strongly-correlated oxide [4, 13, 18, 23, 33]. Moreover, generalisations to DFT + $U$ such as first-principles methods for calculating the Hubbard $U$ parameter [13, 18], the DFT + $U$+$V$ method for including inter-site interactions [34], TDDFT + $U$ [35] and, most pertinent for this study, previous investigations into subspace representations of nonorthogonal Hubbard projectors in DFT + $U$ [7, 8] have also been applied successfully to this system.

Its principal properties of interest, already described in Chap. 2 are that it is a type-II antiferromagnetic insulator with a local magnetic moment of between 1.64 and 1.9$\mu_B$ [18] and an optical gap of approximately 4 eV with predominantly charge-transfer, oxygen 2$p$ to nickel 3$d$, character [36, 37]. We have chosen to study NiO, therefore, because it is so well characterised and we do not seek to revise any previous results. Quite on the contrary, we have performed calculations which we hope will be complementary to those described in Ref. [7], where the "full", "on-site" and "dual" representations of a linear-combination of pseudo-atomic orbital basis were compared in a study on this material.

### *4.4.1 Computational Methodology*

Calculations of the ground-state electronic structure of bulk antiferromagnetic nickel oxide were carried out within collinear spin-polarised Kohn–Sham DFT [9, 38], and the simplified DFT + $U$ method proposed in Ref. [18]. The linear-scaling ONETEP first-principles package, which is described in detail in Refs. [10, 39, 40], was used. The pseudopotential approximation[1] [41] was employed to obviate the explicit treatment of the core states. The LSDA (PZ81) exchange-correlation functional of Ref. [42] was invoked throughout. Periodic boundary conditions were used with a 512-atom supercell in order to provide an adequate equivalent sampling in real-space of the crystal momentum-dependence of the Kohn–Sham eigenstates. A systematic variational basis of Fourier-Lagrange, also known as periodic cardinal sine or *psinc*, functions [43, 44], was used, equivalent to a set of plane-waves bandwidth limited to a kinetic-energy cutoff of 825 eV.

In the ONETEP method, the Kohn–Sham density-matrix is represented in the separable form

$$\hat{\rho}^{(\sigma)}\left(\mathbf{r}, \mathbf{r}'\right) = \phi_\alpha\left(\mathbf{r}\right) K^{(\sigma)\alpha\beta} \phi_\beta\left(\mathbf{r}'\right) \tag{4.40}$$

in terms of a set of covariant nonorthogonal generalised Wannier functions (NGWFs) [10], $\{\phi_\bullet(\mathbf{r})\}$, and a corresponding contravariant density kernel, $K^{\bullet\bullet}$, for each spin channel. The density kernel was untruncated in the calculations described here. In the ONETEP method [39, 40], the total energy is iteratively minimised both with respect to the elements of the density kernel [45] for a given set of NGWFs, while using a combination of penalty-functional [46] and LNVD [47–49] techniques to ensure the validity of the density matrix, and with respect to the expansion coefficients of the NGWFs in the psinc basis. The converged NGWFs (a minimal set of nine functions for nickel $4s$, $4p$, $3d$ and 4 for oxygen $2s$, $2p$, truncated to an atom-centered sphere of 4.0 Å were employed in calculations on NiO) are those which are optimised to minimise the total energy and are thus adapted to the chemical environment, incorporating all valence-electron hybridisation effects in the ground-state density.

Our principal purpose was to provide an appraisal of the difference in predicted electronic properties, if any, given by DFT + $U$ when using nonorthogonal Hubbard projectors with either the "dual" or "tensorial" representations of the correlated subspaces. The "dual" representation, in particular, was selected for comparison since it appears to be the most sophisticated of the previously proposed subspace definitions—although it does break the tensorial invariance of the total energy at the expense of preserving the Hermiticity of the potential, it has a tensorially invariant occupancy matrix trace which cannot be said of the manifestly incomplete "on-site" and "full" representations. The latter three representations were previously compared in detail in Ref. [7].

In order to provide an unbiased analysis, we used the same experimental lattice constant for all calculations and, while it perhaps would be interesting to separately optimise the lattice with the two methods, we leave issues of structural relaxation to a future study. Moreover, in order to obviate intervention in the construction of

---

[1] A set of RRKJ Pseudopotentials were generated using the Opium code, http://opium.sourceforge.net, using the GGA input parameters available therein, optimized for a minimum plane-wave cutoff of 680 eV, albeit with a scalar-relativistic correction for all species and, for the transition-metal ions, some slight modifications to the core radii and a non-linear core correction of Fuchs-Scheffler characteristic radius 1.3 a.u.

the correlated subspaces so far as possible, we carried out the DFT + $U$ calculations in the projector self-consistent formalism described in Chap. 3 and Ref. [1]. We also include, for the purposes of comparison, the results of conventional DFT + $U$ calculations using hydrogenic $3d$-orbital Hubbard projectors for each subspace (in which case there is no ambiguity in the representation for a given projector profile) which were used as the initial guess for the projector self-consistency cycle in both the "dual" or "tensorial" cases.[2]

In the projector self-consistent DFT + $U$ scheme of Ref. [1], as we have described in more detail in chapter [3], the set of five converged NGWFs of maximal $3d$-orbital character on a transition-metal atom responsible for strong correlation effects (or seven converged NGWFs of maximal $4f$-orbital character on a lanthanoid atom, or possibly more exotic converged bond-centered generalised Wannier functions) are selected as Hubbard projectors to redefine the DFT + $U$ occupancy matrices for the total energy minimisation in the next projector iteration. The energy is not directly minimised with respect to the expansion coefficients of the Hubbard projectors (since it would violate the variational principle if either the Hubbard projectors or the interactions $U$ were allowed to change during energy minimisation [14]), but the updated projectors nonetheless converge towards those which equal a subset of the NGWFs which provide the ground state energy corresponding to the DFT + $U$ correction which they themselves define. The projector-update process is repeated, alternating between direct variational minimisation of the total energy for a fixed set of projectors and renewal of the projectors (in a manner reminiscent of the density-mixing method for solving non-linear systems [50, 51]—though no actual mixing from previous projector iterations is needed for numerical stability) until both the density and projectors are individually converged.

Figure 4.1 shows the set of nine converged NGWFs on a nickel atom in NiO (LDA+$U$ = 6eV, "tensorial" representation) at projector self-consistency. Those five with maximal projection onto the "initial guess" hydrogenic $3d$-like orbitals are those which are used to build the correlated subspace for that atom, although there is little doubt in practice as to which are the most localised. The Hubbard projectors have adapted, at convergence, to the crystal (or molecular) environment of their host atom and the associated correlated subspace is usually then somewhat deformed away from spherical symmetry in order to describe electronic hybridisation effects.

It was of necessity to investigate the behaviour of DFT + $U$ under different definitions without the complication introduced by a definition-dependent interaction $U$. While it would be of interest to compute either a scalar or tensor $U$ using different definitions of the subspace projection and one of the sophisticated methods available for this task, we felt that it lay beyond the scope of this work to do so. Thus, we performed conventional DFT + $U$ and projector self-consistent DFT + $U$ calculations across a

---

[2] The effective nuclear charge $Z$ used to construct the hydrogenic projectors, which may influence predicted observables in the case of system-independent Hubbard projectors [1, 13] but which does not significantly influence results at projector self-consistency, was estimated by fitting the hydrogenic radial probability density to that of the corresponding valence pseudo-orbitals in the sense of least squared deviations (resulting in $Z = 9.02$ and $Z = 9.10$ for $3d$-orbital projectors in nickel and copper, respectively).

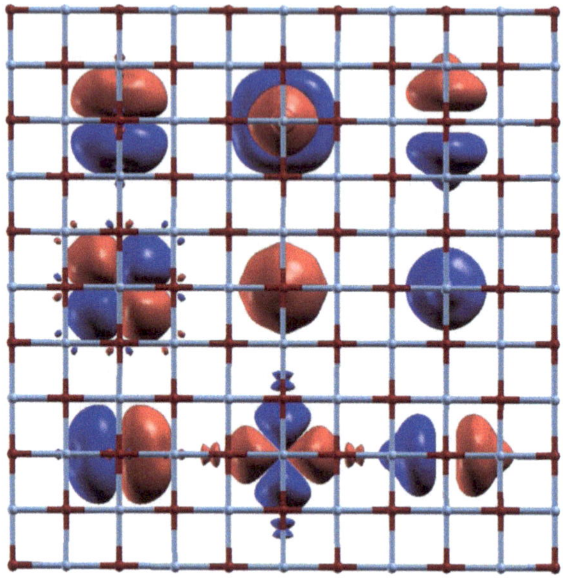

**Fig. 4.1** Isosurfaces of the set of nonorthogonal generalised Wannier functions (NGWFs) on a nickel atom in NiO. The NGWFs are those computed at projector self-consistency in the "tensorial" representation at LDA+$U$ = 6eV. Those in the *left column* (predominantly $3d - t_{2g}$ character) and the *top* and *bottom* NGWFs in the *middle column* (predominantly $3d - e_g$ character) are those used as Hubbard projectors, while the remaining NGWFs (pseudised $2s$-like in the *centre* and pseudised $2p$-like in the *right column*) lie outside the correlated subspace on that atom. The isosurface is set to half of the maximum for the $2s$ and $2p$-like NGWFs and $10^{-3}$ times the maximum for the $3d$-like NGWFs

range of scalar $U$ parameters, which is perhaps best viewed as a parameter controlling the strength of the idempotency penalty-functional on the correlated subspaces since no off-diagonal elements were included in the interaction parameter.

### *4.4.2 Occupancies and Magnetic Dipole Moments*

In agreement with previous studies [7, 8] on NiO, we have found the DFT + $U$ occupancy matrix and local magnetic dipole moment associated with the correlated subspaces in this material depend significantly on the definition used for the correlated subspace projection operator.

Turning first to the total occupancy of the correlated subspaces, shown in Fig. 4.2, we find a steady decrease with increasing $U$ parameter, which is almost entirely due to the DFT + $U$ correction introducing a repulsive potential to the less-than-half occupied nickel $e_g$ orbitals of the minority spin channel. Conversely, we notice that for the largest element on the diagonal of the occupancy matrix (which is almost

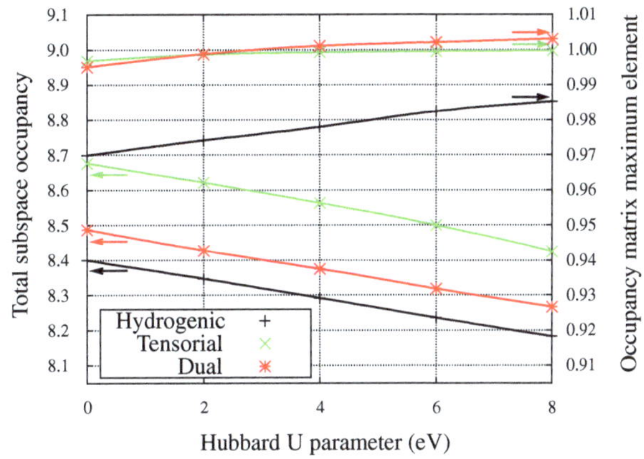


**Fig. 4.2** The total occupancy of a correlated subspace in NiO (*left axis*) and the maximum element on the diagonal of the occupancy matrix (*right axis*), as a function of the interaction $U$. Values are computed with hydrogenic Hubbard projectors and self-consistent NGWF projectors in both the "dual" and "tensorial" representations

identical to that of the other majority orbitals of the same symmetry at a given $U$), DFT + $U$ introduces an attractive potential tending to fully occupy the corresponding orbital.

Based on the data shown on the maximal occupancy element for hydrogenic projectors, for those projectors most commonly used in DFT + $U$ which are not adapted to their chemical environment and so cannot fully account for densities deviating from spherical symmetry, we would conjecture that a rather excessive $U$ value would be needed to complete the orbital filling. In such a case, the attractive DFT + $U$ contribution to the potential is expected to remain spuriously large due to the inadequate description of electronic hybridisation. On the other hand, if we look at self-consistent NGWF projectors in the "dual" representation, there is a tendency to overfill the few most occupied Hubbard projectors, to wit the occupancy begins to exceed unity beyond $U \approx 3$eV. This latter affliction is a rather hazardous one for the DFT + $U$ functional, since the contribution to the energy correction arising from orbitals exhibiting it may become negative in severe cases; this is incorrect behaviour for a penalty-functional. The reason behind this excessive occupancy is the spurious non-locality of the Hubbard projector duals in the "dual" representation, they may pick up density contributions from all across the simulation cell. On the contrary, when self-consistent projectors are used in the "tensorial" representation, the maximal matrix elements tend asymptotically to unity with increasing $U$, as expected (reaching 0.9998 at $U = 8$eV).

Considering the local magnetic moment on the nickel atoms, depicted in Fig. 4.3, we observe the expected increase with the $U$ parameter as the majority and minority channels of the magnetisation-carrying orbitals become increasingly filled or emptied, respectively, as DFT + $U$ enforces the idempotency of the subspace density-matrices more strongly. Beyond a certain value of $U \approx 2$–3eV, the response to perturbations stiffens and following this point the increase in magnetisation is approximately linear with $U$.

The local magnetic dipole moment has been computed in two different ways, one being the difference of the traces of the DFT + $U$ occupancy matrices of the two spin-

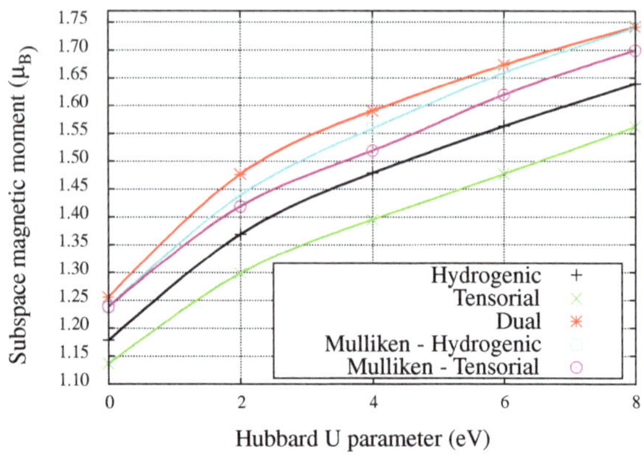


**Fig. 4.3** The local magnetic dipole moment on nickel atoms in NiO as a function of the interaction $U$. Values are computed as in Fig. 4.2. Both the projection of the moment onto the DFT+$U$ correlated subspace and the Hubbard projector independent local moment given by Mulliken analysis are shown

channels (and so explicitly dependent on the definition of the subspace projection), another denoted "Mulliken" corresponding to the total moment associated with the particular atom (and so is not explicitly dependent on the definition of the subspace projection but implicitly via changes to the density) provided by the expression

$$\sum_{\alpha \in \mathcal{A}^{(I)}} \sum_{\beta \in \mathcal{S}} \left( K^{(\uparrow)\alpha\beta} - K^{(\downarrow)\alpha\beta} \right) S_{\beta\alpha}, \tag{4.41}$$

where $S_{\bullet\bullet}$ is the covariant metric on all NGWFs in the simulation cell and $\mathcal{A}^{(I)}$ signifies the set of all NGWFs centred on the host atom of correlated subspace $I$. The two definitions yield similar results in the "dual" case since there is minimal magnetisation carried by the less localised NGWFs on the nickel atoms. The observed behaviour is that the NGWF projectors in the "dual" representation yield higher local moments than the representation-independent hydrogenic projectors and, moreover, that yielded by the latter is larger than that from NGWF projectors in the "tensorial" representation. Consequently, we would expect the exchange splitting between energy levels, which makes up a large contribution to the insulating gap in this material (it is well-described within unrestricted Hartree–Fock theory [36]), to follow the same trend. While this behaviour may seem a somewhat unfavourable reflection on the "tensorial" representation, it is fully in line with our understanding that the "dual" representation (or any related delocalised "Mulliken"-type analysis) picks up spurious contributions from neighbouring atoms' magnetisation densities by construction. The difference between the "tensorial" and "Mulliken-Tensorial" curves, $\approx$0.10–0.15$\mu_B$, is precisely this erroneous non-local contribution. The discrepancy between the "hydrogenic" and "tensorial" curves is of a different nature, since both are made up of subspace-local moments only, and it is both explicitly dependent on differences in the spatial profile of the projectors and implicitly dependent due to the consequent changes to the electronic density.

Finally, a comment on Fig. 4.4, in which the magnetisation density in the through-centers (001) plane is shown at LDA+$U$ = 6eV with hydrogenic Hubbard projectors along with its change upon moving to self-consistent NGWF projectors in both

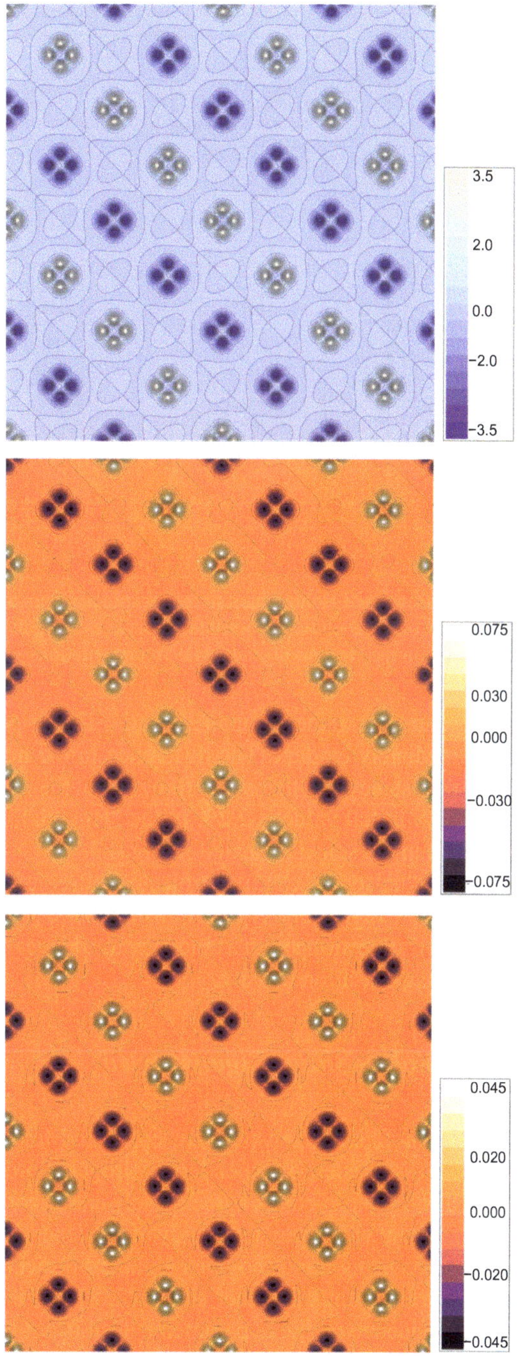


**Fig. 4.4** Magnetisation density of NiO, at LDA+$U = 6$eV using hydrogenic Hubbard projectors, in the (001) plane through atomic centres (*top*). The change upon the aforementioned which is yielded via self-consistency over NGWF Hubbard projectors, in the "tensorial" (*middle*) and "dual" (*bottom*) representations, is also shown. Regions of high magnetisation density correspond to nickel ions

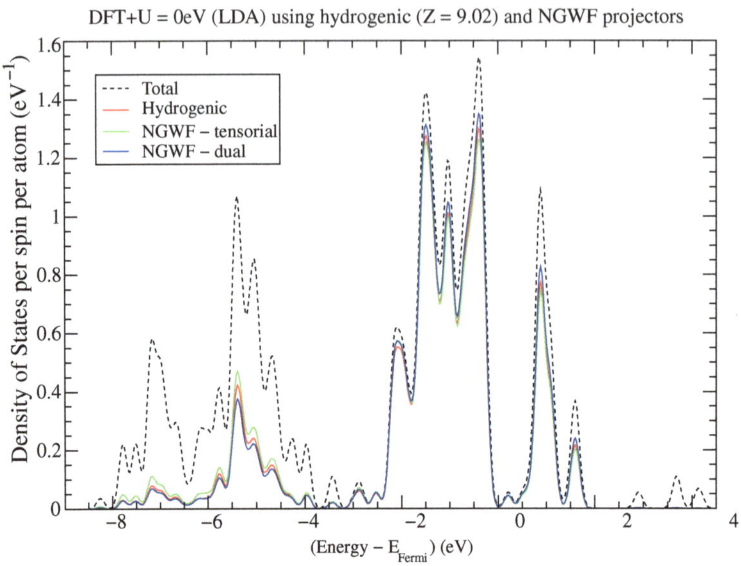


**Fig. 4.5** Density of Kohn–Sham states per spin per atom of NiO at LDA+$U$ = 0eV together with its projection onto the union of correlated subspaces using hydrogenic Hubbard projectors and NGWF projectors in the "tensorial" and "dual" representations

representations of the occupancy matrix. The striped magnetism in the (111) direction is immediately evident. The difference in the magnetisation provided by projector self-consistency is rather small (the ratio of the whole-cell RMS change to the whole-cell RMS initial value is 2.2% in the "tensorial" case and 1.2% in the "dual" case) and acts in the sense indicated in Fig. 4.3.

In the "tensorial" case the projector self-consistency acts to suppress the magnetisation in the region of oxygen atom centres very slightly, while the opposite occurs in the "dual" case along with a tendency to add weak nickel-centered radial oscillations in the magnetisation density.

### 4.4.3 Kohn–Sham Eigenspectra

The Kohn–Sham eigenspectrum computed for NiO using DFT + $U$ with both our "best guess" system-independent hydrogenic projectors and self-consistently determined NGWF projectors agree closely and accord with previous studies. Moreover, in agreement with a previous study of the dependence on the occupancy matrix definition when using nonorthogonal Hubbard projectors, the representation dependence of spectral features is rather subtle [7, 8] and is considerably less significant than the dependence on the $U$ parameter. That is not to say, however, that the differences yielded may be guaranteed to be fully recovered by a self-consistent determination or arbitrary variation of the interaction $U$, since we observe different dependences on this parameter for different spectral peaks depending on the correlated subspace representation.

In order to gain insight into the spectral decomposition of the DFT + $U$ projection operator for a fixed ground-state density, we begin with the density of Kohn–Sham states, and its projection onto the union of correlated subspaces, at LDA+$U$ $= 0\,$eV (at which value there is no projector-dependence in the total spectrum and no representation-dependence on the NGWF Hubbard projectors). This, illustrated in Fig. 4.5, shows the energy interval in which the DFT + $U$ takes explicit effect. All three subspace definitions have very similar projector weight from approximately 4 eV below to 5 eV above the Fermi level (which is set to the mid-point of the insulating gap; a Gaussian smearing half-width of 0.1 eV is employed). At energies below this, principally corresponding to states of majority-channel nickel $3d$, but also with a substantial oxygen $2p$, character, the "tensorial" representation of NGWF projectors has a somewhat greater weight than the hydrogenic and "dual" representations, although the differences are not very substantial.

At conduction-band energies above 5 eV, we expect and observe that the correlated subspaces have increasingly negligible projection towards higher energies (not visable on a linear scale). The hydrogenic projectors, which are unadapted to the electronic structure of the system, exhibit a persistent, very weak weight which extends well into the conduction band. The self-consistent NGWF projectors, being better suited to their chemical environment, exhibit less of this small spurious energetic non-locality and the decay with energy is more rapid in the "tensorial" than in the "dual" case. We hasten to add that the effect is rather weak and that, in any case, Hubbard projectors in the form of orthonormal Maximally-Localised Wannier functions [30, 31] (particularly those self-consistently determined with the DFT + $U$ eigenstates), would allow for explicit truncation in the energy range, although perhaps at the expense of having a greater spatial extent.

Considering, next, a Hubbard $U$ value within the range of values known to give reasonable agreement with experiment, namely $U = 6$eV, we have shown the total density of states (DoS), and its correlated subspace projection, in the three representations of interest, in the top panel of Fig. 4.6. The bottom panel shows the decomposition of the "tensorial" DoS into its contributions from oxygen atoms and both predominantly spin-aligned (majority) or spin-antialigned (minority) nickel atoms. Although all of the dominant features are shared between the eigenspectra of the various representations, there are some discrepancies which are worthy of mention. Most notable is the trend for the insulating gap (see also Fig. 4.7) to open slightly, predominantly in the minority $e_g$ peak at $\approx 2\,$eV, as we go from "tensorial" NGWF (2.35 eV) to hydrogenic (2.60 eV) to "dual" NGWF representations (2.68 eV). We attribute this primarily to changes in the exchange splitting provided by the enhancement of the local magnetic moment, which follows the same trend, as can be seen in Fig. 4.3. We also note a slight suppression of the localised character of the valence band edge following the same trend, corresponding to states known to have a mixed nickel $3d$–oxygen $2p$ character. Lower in energy at approximately $-7.5$eV, the amplitude of the majority $e_g$ peak is somewhat suppressed in the "tensorial" representation

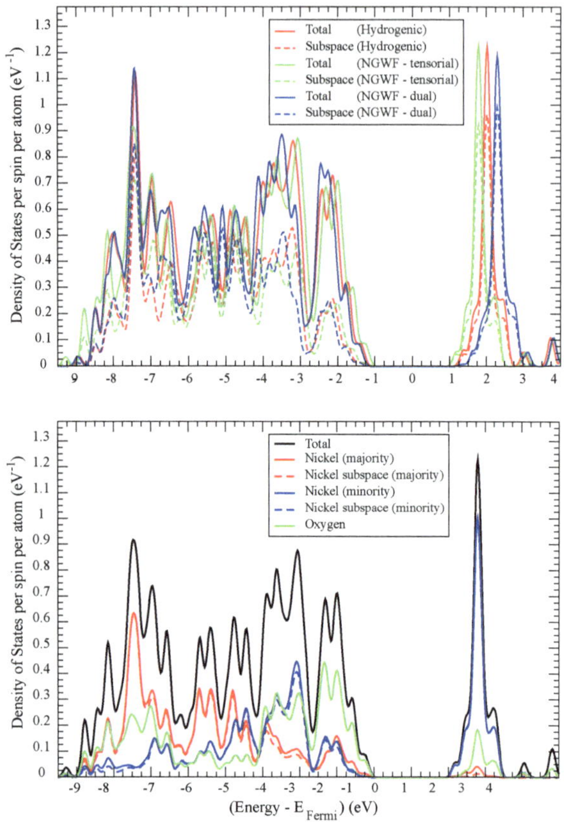


**Fig. 4.6** Density of Kohn–Sham states per spin per atom of NiO at LDA+$U$ = 6eV, together with its projection onto the union of correlated subspaces using hydrogenic Hubbard projectors and NGWF projectors in the "tensorial" and "dual" representations (*top*). Also shown (*bottom*) is the decomposition of the NGWF-"tensorial" density of states for a given spin channel into its contributions from NGWFs on nickel atoms with magnetisation aligned (majority) and anti-aligned (*minority*), the correlated subspace projections of each, and the contribution due to oxygen NGWFs

in favour of a broader shoulder of strongly hybridised character towards lower energies of around $-9$eV.

Lastly, for NiO, we show the $U$-dependence of the total DoS in the three DFT + $U$ correlated subspace definitions in Fig. 4.8 and the Kohn–Sham insulating gap in Fig. 4.7. In all cases we recover the canonical DFT + $U$ behaviour for this material, that is with increasing interaction parameter $U$ the tendency for: the low-energy (primarily majority-channel $e_g$-like) peak falls deeper into the valence band as an attractive potential is applied to fill it completely; the strongly nickel $t_{2g}$-like valence-band edge at the LDA level gives way to hybridised oxygen $2p$ character as the $t_{2g}$-like states are pushed to lower energies; and the minority-channel nickel $e_g$-like first peak in the conduction band is increased in energy as it partial occupancy is subjected to repulsive potential corrections.

The "dual" representation provides a rather similar eigenspectrum to the conventional DFT + $U$ method when using unoptimised hydrogenic Hubbard projectors, particularly in the line-shape of the conduction band, although the Kohn–Sham gap is marginally broader and the majority $e_g$ states deep in the valence band have a greater amplitude. The "tensorial" representation also provides very similar spectra,

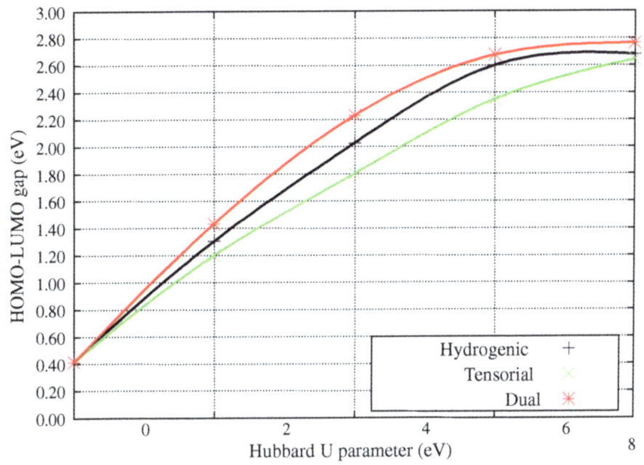


**Fig. 4.7** The Hubbard $U$ dependence of the Kohn–Sham band-gap of NiO at LDA+$U$, computed as in Fig. 4.7

but the gap is closed somewhat with respect to the hydrogenic model, as discussed. A noteworthy difference between the spectra of the "tensorial" representation and the alternatives is that, generally, both the majority and minority $e_g$ peaks lie slightly lower in energy and the energy difference between the two (the spin-flip excitation) is somewhat reduced. Overall, we reiterate that the effects on the spectra due to the local or non-local construction of the Hubbard projector duals are not sufficiently great to reasonably draw conclusions regarding the relative merit of methods based on agreement, or otherwise, with experimental observations. Rather, in this matter, points of principle such as the preservation of tensorial invariance, or the avoidance of occupancies exceeding unity, must therefore take precedence in our view.

## 4.5 Magnetic Molecule: The Copper Phthalocyanine Dimer

Open-shell molecular systems containing transition metal ions sometimes pose a challenge to *first principles* simulation within LDA-based approximations [52]. This is partially due to the tendency of such approximate exchange-correlation functionals to excessively delocalise magnetisation-carrying orbitals in such systems. As noted in Refs. [21, 53–55], both energetic properties such as magnetic exchange coupling and also spectroscopic features such as the nature of the insulating gap and multiplet splittings can consequently be poorly reproduced by such functionals. Sophisticated ab initio techniques such as the $GW$ approximation and local correlation methods such as DFT + $U$, whose traditional realm of application lies in extended systems such as extended oxides and their interfaces, are being increasingly applied to molecular systems and clusters, see for example Refs. [21, 53, 55–57] and references therein.

Spatially-localised corrections such as DFT + $U$ and its variants are inherently short-ranged and fit rather seamlessly into linear-scaling implementations of DFT, in which case it is then often more natural to employ a nonorthogonal set of Hubbard projectors [7, 10–12]. It is thus of some importance, and perhaps timely, to consider

**Fig. 4.8** The Hubbard *U* dependence of the Kohn–Sham density of states per spin per atom states of NiO at LDA+*U* as calculated using hydrogenic Hubbard projectors (*top*) and NGWF projectors in the "tensorial" (*middle*) and "dual" representations (*bottom*)

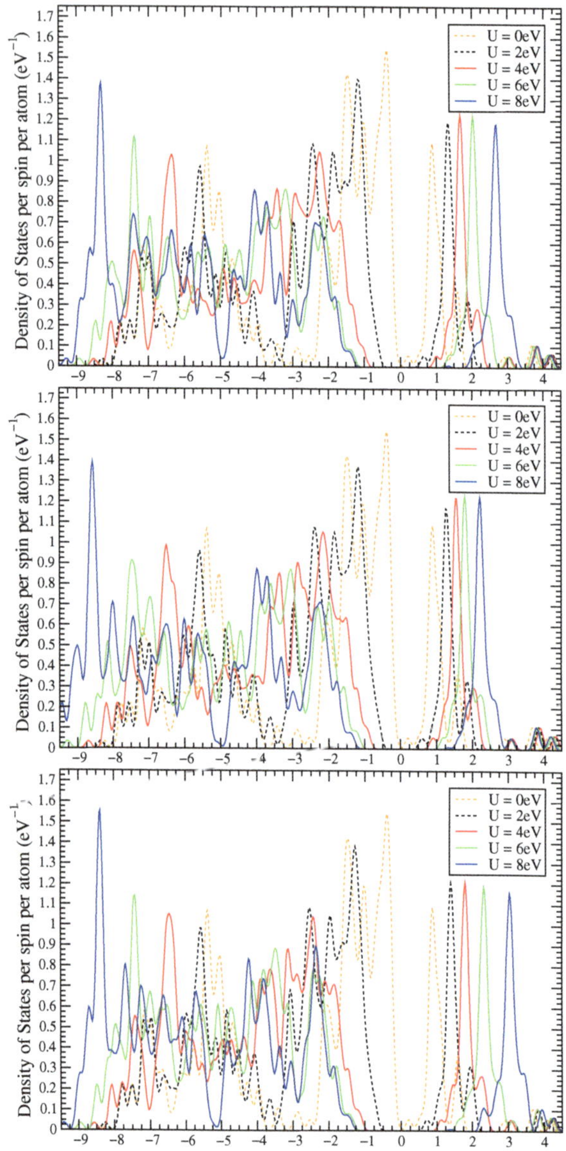


molecular systems on a similar footing to solids when considering the merit of projection methods for DFT + *U*. The correlated orbitals in molecular systems may be rather more spatially diffuse and deviate further from spherical symmetry than their counterparts in solids. As a result, the issue of Hubbard projector-dependence in DFT + *U* and then the manner in which the projection operator is constructed from those projectors, particularly the degree of non-locality in the Hubbard projector duals, can be expected to play a more significant role in the description of molecular systems with DFT+$U$ than is the case in more tightly packed structures.

With a view to analysing the dependence on the correlated subspace definition, or occupancy representation, in the case of molecular systems, we applied our methodology to the ground-state of a binuclear open-shell singlet (antiferromagnetically coupled) complex, the copper phthalocyanine dimer denoted Cu(II)$Pc_2$. Crystalline CuPc is a semiconducting blue dye which, in pure thin-film form and more exotic derivatives, is currently attracting intense experimental and theoretical interest due to its potential for use as a flexible organometallic photovoltaic material [58], as part of field-effect transistors [59] and, due to its magnetic functionality, in spintronic data storage or processing devices [60]. In this system, two correlated subspaces delineated by copper 3$d$-like states are spatially well separated (with approximately 3.77Å between centres) and there is minimal direct electronic bonding between the localised orbitals in the open Cu-3$d$ shells in the two (approximately) planar moeities. The result is a very weak indirect-exchange (i.e., acting via intermediary delocalised ligand states) $S = \frac{1}{2}$ antiferromagnet with a Heisenberg exchange coupling constant of $J \approx -1.50K$; for a detailed analysis of this mechanism we refer the reader to Ref. [61].

## 4.5.1 Computational Methodology

A set of 9 NGWFs (4$s$, 4$p$ and 3$d$) were used for transition-metal ions, 4 each for carbon and nitrogen (2$s$ and 2$p$) and 1 for hydrogen (1$s$). A large NGWF cutoff radius of 5.3 Å and a kinetic-energy cutoff of 1,000 eV was used. The spin-polarised PBE [62] generalised-gradient exchange-correlation functional was employed. An un-solvated and hydrogenated gas-phase dimer model was extracted [3] from the $\alpha(+)$ Cu(II)$Pc_2$ polymorph structure, with a stacking angle of 65.1 degrees and a distance between molecular planes of 3.42 Å, giving a lateral offset of 1.58 Å, as reported from transmission electron diffraction analysis described in Ref. [63]. While the phthalocyanine dimer systems are of some experimental interest in and of themselves [64–66], and similar metal-phthalocyanine dimers have been recently shown to take on structures which are strongly influenced by dispersion interactions [67] which are beyond the scope of this study, the dimer system is for us primarily a convenient simple model of the antiferromagnetic coupling in the extended solid and thus the unoptimised experimental bond lengths were retained. A simulation cell of 30Å × 30Å × 20Å provided an interatomic spacing between periodic images of at least 13.5Å in plane and 16.5Å out of plane.

[3] We acknowledge and thank Nina Kearsey for her kind provision of the extracted dimer structure of Ref. [63].

### 4.5.2 Magnetic Dipole Moments

The open-shell singlet fragments of the Cu(II)$Pc_2$ system consist of single spins on each copper centre, antiparallel with respect to each other. As such, for the exact exchange-correlation functional, which contains a derivative-discontinuity with respect to the occupancies of the exact magnetisation-carrying orbitals, in combination with a population analysis which exactly reproduces the form of those singly-occupied orbitals, the local magnetic moment would be identically $1\mu_B$. However, since an approximate exchange-correlation functional may lower the energy by delocalising and partially occupying those orbitals [52], a significantly lower value was recovered. The DFT + $U$ method seeks to ameliorate this condition in two complementary ways, that is by introducing a derivative-discontinuity to the energy functional which penalises fractional occupancies of the spin-orbitals defined by the subspace projections and also, in doing so, by effectively constraining the Kohn–Sham spin-orbitals to more closely resemble the (usually more localised) spatial form of the correlated subspace.

In light of this, the correlated subspace projected magnetic dipole moments, shown in Fig. 4.9, indicate that the DFT + $U$ method does not effectively localise the magnetisation density to the copper $3d$ manifold for any reasonable value of the $U$ parameter when a localised subspace projection is used. Using conventional hydrogenic Hubbard projectors, with our best guess for the radial profile, we see that there is only a very slight increase in the local moment with $U$. Switching to self-consistent projectors in the "tensorial" representation, we find that it is effectively $U$-independent and reduced with respect to the hydrogenic result, though very slightly so since these projectors differ only very slightly in form from the hydrogenic initial guess in this particular system.

We have included, for completeness, the values yielded by the approximate "truncated dual" representation, which yielded spurious occupancies greater than unity for every value of $U$ considered. Conversely, the "dual" representation yields a greater magnetic moment than the "tensorial" representation, by approx. $0.1\mu_B$ at $U = 0$eV, and this increases steadily at a rate of $\approx 0.02\mu_B \text{eV}^{-1}$. The reason for this discrepancy is readily understood via the atom-decomposed Mulliken analysis of Eq. 4.41. This gives $0.10 - 0.12\mu_B$ on each nitrogen atom which is a nearest-neighbour to copper, as is visualised in the spin-density isosurfaces of Fig. 4.10, and this value remains approximately constant irrespective of the representation or the $U$ parameter.

In spite of their adaptation to the molecular environment, the self-consistent NGWF projectors remain predominantly on the home copper ion and do not have sufficient weight on the neighbouring in-plane nitrogen $2p$ orbitals to capture the magnetisation density associated with them. As a result, in the same manner as the conventional projectors, they fail to retrieve the magnetisation to the copper $3d_{x^2-y^2}$ orbital within DFT + $U$. The "dual" representation, however, overcomes this obstacle apparently due to the dual Hubbard projectors extending over all of the delocalised states in the system, most notably the other regions which contribute to the magnetisation density.

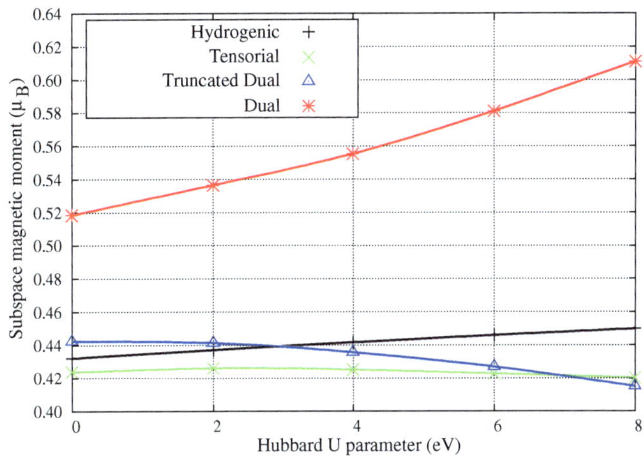


**Fig. 4.9** The average magnitude of the projection of the magnetic dipole moment onto the correlated subspaces of Cu(II)$Pc_2$, plotted as a function of $U$ for various definitions of the subspace projection

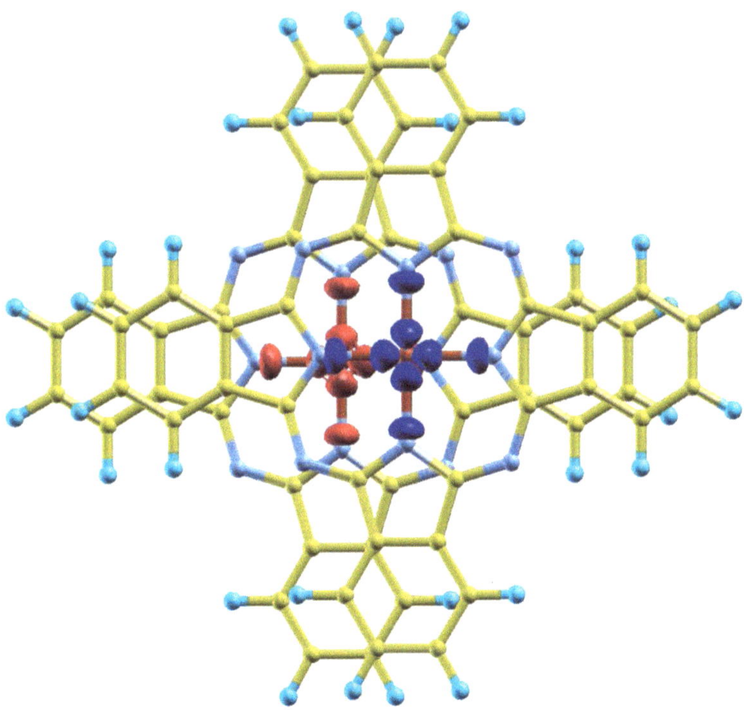

**Fig. 4.10** Spin-density isosurfaces at 5% of maximum in Cu(II)$Pc_2$ at projector self-consistent GGA+$U$ = 6eV with the "tensorial" subspace representation

### 4.5.3 Kohn–Sham Eigenstates

The currently accepted understanding [61, 68, 69] of the spectroscopic nature of the gap in the copper phthalocyanine monomer is that the HOMO level is dominated by a doubly-occupied $a_{1u}$ orbital (each moiety is approximately fourfold $D_{4h}$ symmetric) which consists of a superposition of carbon $p_z$ orbitals delocalised on the pyrrole rings of both monomer units, while the spectroscopically correct LUMO level is a also a delocalised doubly-degenerate orbital, of $e_g$ symmetry composed of a superposition of $\pi$ orbitals on pairs of macrocycle carbon atoms. We do expect some minor enhancement of the gap and perhaps alteration of its spectroscopic nature in the dimer system with respect to the monomer, primarily due to $\sigma$-bonding between delocalised orbitals on either moiety, but for the main features of the electronic structure to be preserved.

It has previously been shown that LDA and GGA-type exchange-correlation functionals do not correctly reproduce the qualitative ordering of states close to the Fermi level in this system, a pathology which has been attributed to self-interaction errors in such functionals [68, 70, 71]. The DFT + $U$ insulating gap of the dimer system within various representations is shown in Fig. 4.11, along with the $U$-dependence of the states nearest the Fermi energy. For the spin-polarised PBE functional we find a small gap of 0.7eV for the dimer, whose nature is a charge-transfer excitation between $b_{1g}$ orbitals on either moiety. The $b_{1g}$ orbital is that which carries the magnetisation density in the dimer, consisting primarily of copper $3d_{x^2-y^2}$ $\sigma$-bonded to in-plane nitrogen $2p$, and as such this system is analogous to the Mott-Hubbard insulating solids usually treated with the DFT + $U$ correction where both the valence and conduction-band edges are directly probed by the Hubbard projectors.

The representation dependence of the HOMO–LUMO gap follows the same trend as the local magnetic moment, due to the DFT + $U$ correction to the Coulomb-repulsion gap being somewhat augmented by an associated enhancement to the exchange splitting caused by the increase in the local magnetic moments. In the case of the HOMO orbital, a small value of $U$ is needed to push the singly-occupied $b_{1g}$ state to its spectroscopically correct position below the $a_{1u}$ state and the effect is rather more strongly pronounced in the "dual" representation than in the spatially-localised methods. The "tensorial" and "hydrogenic" representations have similar effects, as expected, since the effect of projector self-consistency is rather small in this system. In the case of the virtual orbitals, the localised $b_{1g}$ character of the LUMO persists for the "tensorial" and "hydrogenic" methods, which behave similarly, while $U \geq 6$eV is sufficient to expose a delocalised $e_g$ orbital as LUMO in the "dual" representation. There is necessarily some small perturbative effect on delocalised orbitals induced by changes to those which are DFT + $U$ corrected. This is evident in all projection techniques, however it is worth noting that the effect is significantly stronger for the "dual" representation, a result of those states having a larger projection onto the correlated subspaces in that representation.

The overall result is that for this strongly-hybridised system, the "dual" representation recovers the expected magnetic dipole moment with significantly more

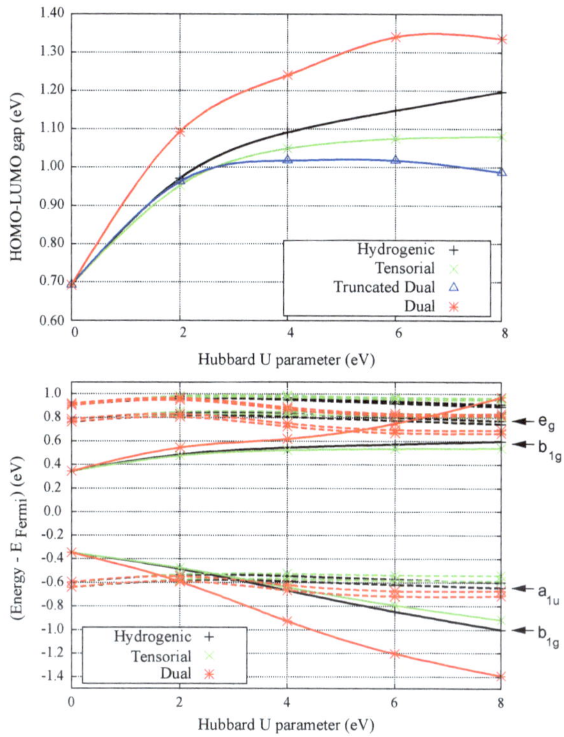


**Fig. 4.11** The HOMO–LUMO energy gap (*top*) and the energy levels adjacent to the Fermi energy (*bottom*) of Cu(II)$Pc_2$, plotted as a function of $U$. *Solid lines* show energy levels of states of predominantly Cu-centered $b_{1g}$ character, and to which the DFT+$U$ correction strongly applies, while *dashed lines* show energy levels of states of more delocalised nature (*bottom*)

success than the fully localised projections. The spectroscopic nature of the insulating gap is also recovered to a greater degree for a given value of $U$. We would contend, however, that it does so for reasons not expected in the DFT+$U$ method. Specifically, where the local magnetic moment as measured by the "dual" projectors increases with increasing $U$, the spatial distribution of this increase is dominated by the region surrounding the copper ion but also may have substantial diffuse contributions, as opposed to the "tensorial" or orthonormal "hydrogenic" contributions with which we are guaranteed to include only subspace-localised densities. Quite aside from the issue of the broken tensorial invariance inherent to the "dual" representation, it introduces additional complexity to the already delicate task of selecting the Hubbard projectors since its occupancy matrices depend on the nature of all of the basis functions in the simulation cell and their overlaps.

It is clear that in order to capture the diffuse magnetisation densities in this system with DFT+$U$ while retaining the tensorial invariance of the total energy, it would be necessary to include additional in-plane nitrogen $2p$-like Hubbard projectors in the spanning-set of correlated subspaces in the "tensorial" representation (copper $3d$ and nitrogen $2p$ function overlap disallows the "hydrogenic" representation). In such an approach, however, it would seem overly simplistic to use a single scalar $U$ parameter for each subspace and a more sophisticated projector-decomposed $U$ tensor would be more appropriate.

## 4.6 Concluding Remarks

We have presented a revised formalism for the construction of projection operators, and consequently the occupancy matrices, of strongly-correlated subspaces using nonorthogonal Hubbard projector functions in ab initio methods such as DFT + $U$ and DFT+DMFT. In contrast to the previously proposed "full" [13], "on-site" [14] and "dual" [7] representations, our definition preserves the important property of tensorial invariance in the total occupancy of each subspace, the total energy and the ionic forces, by construction. The expressions needed to implement a tensorially-invariant DFT + $U$ energy functional and the resulting potential and ionic forces, both with scalar and tensor interaction parameters, have been presented, along with the adaptations needed to integrate the formalism with a well-known method for computing the interaction $U$ from first-principles [18].

It is frequently the case that a subset of the localised nonorthogonal basis functions that are used to represent the Kohn–Sham orbitals are also used as Hubbard projectors, in practice, either for computational convenience or to achieve projector self-consistency [1]. In both an analytical study and in a numerical study of a solid-oxide and a correlated molecular system, we have shown that it may be somewhat inappropriate to continue to identify the dual space and the metric tensor of the basis functions with the dual space of Hubbard projectors on each site. For molecular systems, in particular, the discrepancy compared to orthogonal projectors which is thereby introduced may be significant. The resulting projector duals (contravariant vectors) are unsuited to constructing a correction for localised correlation effects, being spatially delocalised across the entire simulation cell in general. Consequently, with delocalised duals, a tensor-incompatible symmetrisation of the projection operator is needed to ensure a Hermitian potential. This may result in unphysical occupancy matrix elements and an uncontrolled spatially delocalised action of the corrective potential which it defines. Put simply, additional non-local interaction terms are introduced in the "dual" approximation which are extraneous to the requirement of accounting for the nonorthogonality of the Hubbard projectors.

Our tensorial formalism may be implemented in any methodology which makes use of a nonorthogonal set of functions to define each correlated subspace and it inherently retains the spatial localisation of Hubbard projector duals if the projectors are themselves localised. Consequently, it is less computationally expensive and simpler to implement in linear-scaling methods in practice than the "on-site" or "dual" representations which employ delocalised dual projectors. No spurious contributions to the occupancy of a given subspace are assigned from the complement of that subspace in the proposed formalism, in contrast to the aforementioned formalisms and, as a result, the corrective potential arising from it is automatically Hermitian and localised to the same subspace.

It is our hope that we have dispelled some of ambiguities surrounding this topic which we feel have arisen inevitably as a result of the neglect of the invaluable tensor notation. Moreover, as the use of linear-scaling ab initio approaches becomes increasingly widespread, we envisage that this work may aid the routine implementation of sophisticated functionality in the nonorthogonal bases typically employed, obviating the expenditure of explicit orthonormalisation.

## References


1. D.D. O'Regan, N.D.M. Hine, M.C. Payne, A.A. Mostofi, Projector self-consistent DFT+U using nonorthogonal generalized Wannier functions. Phys. Rev. B **82**(8), 081102 (2010)
2. D.D. O'Regan, M.C. Payne, A.A. Mostofi, Subspace representations in ab initio methods for strongly correlated systems. Phys. Rev. B **83**(24), 245124 (2011)
3. V.I. Anisimov, J. Zaanen, O.K. Andersen, Band theory and Mott insulators: Hubbard U instead of Stoner I. Phys. Rev. B **44**(3), 943 (1991)
4. V.I. Anisimov, I.V. Solovyev, M.A. Korotin, M.T. Czyżyk, G.A Sawatzky, Density-functional theory and NiO photoemission spectra. Phys. Rev. B **48**(23), 16929 (1993)
5. V.I. Anisimov, A.I. Poteryaev, M.A. Korotin, A.O. Anokhin, G. Kotliar, First-principles calculations of the electronic structure and spectra of strongly correlated systems: dynamical mean-field theory. J. Phys. Condens. Matter **9**(35), 7359 (1997)
6. A.I. Lichtenstein, M.I. Katsnelson, Ab initio calculations of quasiparticle band structure in correlated systems: LDA++ approach. Phys. Rev. B **57**(12), 6884 (1998)
7. M.J. Han, T. Ozaki, J. Yu, *O(N) LDA+U* electronic structure calculation method based on the nonorthogonal pseudoatomic orbital basis. Phys. Rev. B **73**(4), 045110 (2006)
8. C. Tablero, Representations of the occupation number matrix on the LDA/GGA+U method. J. Phys. Condens. Matter **20**(32), 325205 (2008)
9. W. Kohn, L.J. Sham, Self-consistent equations including exchange and correlation effects. Phys. Rev. **140**(4A), A1133 (1965)
10. C.-K. Skylaris, A.A. Mostofi, P.D. Haynes, O. Diéguez, M.C. Payne, Nonorthogonal generalized Wannier function pseudopotential plane-wave method. Phys. Rev. B **66**(3), 035119 (2002)
11. E. Hernández, M.J. Gillan, Self-consistent first-principles technique with linear scaling. Phys. Rev. B **51**(15), 10157 (1995)
12. F. Mauri, G. Galli, Electronic-structure calculations and molecular-dynamics simulations with linear system-size scaling. Phys. Rev. B **50**(7), 4316 (1994)
13. W.E. Pickett, S.C. Erwin, E.C. Ethridge, Reformulation of the *LDA+U* method for a local-orbital basis. Phys. Rev. B **58**(3), 1201 (1998)
14. K.K.H. Eschrig, I. Chaplygin, Density functional application to strongly correlated electron systems. J. Solid State Chem. **176**(2), 482 (2003)
15. P.O. Löwdin, On the non-orthogonality problem connected with the use of atomic wave functions in the theory of molecules and crystals. J. Chem. Phys. **18**, 365 (1950)
16. E. Artacho, L.M. del Bosch , Nonorthogonal basis sets in quantum mechanics: representations and second quantization. Phys. Rev. A **43**(11), 5770 (1991)
17. C.A. White, P. Maslen, M.S. Lee, M. Head-Gordon, The tensor properties of energy gradients within a non-orthogonal basis. Chem. Phys. Lett. **276**(1–2), 133 (1997)
18. M. Cococcioni, S. de Gironcoli, Linear response approach to the calculation of the effective interaction parameters in the *LDA+U* method. Phys. Rev. B **71**(3), 035105 (2005)
19. R.S. Mulliken, Electronic population analysis on LCAO-MO molecular wave functions I. J. Chem. Phys. **23**, 1833 (1955)
20. A. Einstein, Die grundlage der allgemeinen relativitätstheorie. Ann. Phys. **354**, 769 (1916)
21. H.J. Kulik, M. Cococcioni, D.A. Scherlis, N. Marzari, Density functional theory in transition-metal chemistry: a self-consistent Hubbard *U* approach. Phys. Rev. Lett. **97**(10), 103001 (2006)
22. P.W. Anderson, Self-consistent pseudopotentials and ultralocalized functions for energy bands. Phys. Rev. Lett. **21**(1), 13 (1968)

69. F. Evangelista, V. Carravetta, G. Stefani, B. Jansik, M. Alagia, S. Stranges, A. Ruocco, Electronic structure of copper phthalocyanine: an experimental and theoretical study of occupied and unoccupied levels. J. Chem. Phys. **126**(12), 124709 (2007)
70. H. Vázquez, P. Jelínek, M. Brandbyge, A. Jauho, F. Flores, Corrections to the density-functional theory electronic spectrum: copper phthalocyanine. Appl. Phys. A **95**, 257 (2009)
71. A. Calzolari, A. Ferretti, M.B. Nardelli, Ab initio correlation effects on the electronic and transport properties of metal(II)phthalocyanine-based devices. Nanotechnology **18**(42), 424013 (2007)

# Chapter 5
# Geometric Aspects of Representation Optimisation

The use of non-orthogonal basis sets for the single-particle electronic wave-function is ubiquitous in linear-scaling density- functional theory and tight-binding approaches, indeed it is vital to exploit the spatial compactness available to the elements of such a basis in order to achieve linear-scaling effort with system size [1–4].

The nonorthogonality of such sets does require a particular care with notation, however. This necessity is, unfortunately, all too often ignored and one occasionally observes basis overlap matrices appearing in a somewhat *post hoc* fashion in expressions. On the contrary, we have found that if one is willing to spend some time acclimatising to the notation of curved spaces, it eventually becomes more natural to assume nonorthogonality in most derivations and to use the index positions as a guide in times of uncertainty.

This chapter is intended both as an introduction to some useful, prerequisite concepts and as an exposition of an original technique, based on geometric principles, for accelerating and improving the robustness of total-energy minimisation in linear-scaling DFT. While we do not expand upon earlier textbook results, all non-standard derivations are provided in full detail. Although, admittedly, these formulae may appear to be a little daunting at a first glance, they really amount to little more than linear-algebra and differential calculus.

## 5.1 Motivation

Many ab initio methods for computing the properties of atomic ensembles make use of a set of non-orthogonal basis functions to express the electronic wave-function. In the case of linear-scaling density functional theory [1–5], these are usually, but not necessarily, spatially-truncated and atom-centred functions which possess the symmetries of atomic orbitals and are defined in direct space. We denote the set of these functions as

$$\{\phi_\alpha(\mathbf{r}) \mid \alpha \in \{1, \ldots, N\}\}, \tag{5.1}$$

D. D. O'Regan, *Optimised Projections for the Ab Initio Simulation of Large and Strongly Correlated Systems*, Springer Theses, DOI: 10.1007/978-3-642-23238-1_5,

where the co-ordinate space on which these functions are defined is represented by **r** and N is the cardinality of the function set. Technically, if the functions $\{\phi_\alpha\}$ are fixed throughout the calculation then these may be described as a basis. In general, however, we allow for the possibility that these functions may change during the course of a calculation and so the co-ordinate vectors **r** strictly form the *basis*, and the representation vectors $\{\phi_\alpha\}$ are usually described as the *support functions* [3].

Of particular interest, of course, are procedures where the support functions are optimised in order to minimise the total-energy, as might by applicable to static density-functional theory [6, 7]. The support functions are described as non-orthogonal generalised Wannier functions [1], NGWFs, in this case. Other possibilities, for example, are where the support functions evolve in time, as the case might be for time-dependent density-functional theory (TDDFT) [8], or indeed where they evolve in order to satisfy some other physically-motivated criterion such as maximising the Coulomb interaction by some measure, as mentioned in Chap. 7.

In this chapter, we concentrate on the first of these possibilities, where the total-energy is minimised, however many of the formulae below will carry directly over to the other situations. Energy minimisation with respect to the support functions (NGWFs) is a central feature of the ONETEP linear-scaling DFT code [9, 10], where it is carried out using a particular non-linear conjugate gradients scheme, though similar energy minimisation methods have been previously employed [4, 11–13] and the general principle is sure to attract further investigation in the future.

In ONETEP, as we have described in Chap. 1, the optimal representation of the Kohn–Sham density-matrix in terms of the support functions, that known as the density kernel, is located (for a given fixed set of support functions) before the gradient of the energy with respect to the support functions (assuming a fixed tensor representation of the density-matrix) is computed. The support functions may be updated using this gradient in different ways, depending on whether a linear or non-linear (quadratic or cubic) Fletcher–Reeves conjugate gradients step is required. Usually, no matter how small the step length, the Hamiltonian matrix is completely re-evaluated when the support functions change, and generally the density kernel is re-purified (made idempotent with respect to the new support functions) but not re-optimised for minimal energy during the conjugate gradients trial step.

Certain questions naturally arise concerning such a method, regarding the nature of the density kernel in particular:

- Supposing we evaluate a change vector to the support functions (with a fixed density kernel corresponding to the optimal density-matrix for the support functions at which the change is evaluated), then how much of the effort spent optimising the density kernel afterwards will just involve recovering something approaching the quality of the original density-matrix?
- Can some of that effort be obviated by automatically correcting the density kernel so that it takes into account the curvature (in the sense of differential geometry) of the Riemannian manifold associated with the support functions and their duals? How much faster would such a method be, and would there be any other advantages associated with it?

- Supposing such a correction could be made, would it imply anything about the preservation of the normalisation, idempotency, or commutativity between the density kernel and the Hamiltonian, all of which are usually destroyed when the support functions change?
- Would it be better to apply such a correction only for the update step actually selected by the nonlinear conjugate-gradients scheme, or to apply it also every time a trial change in the support functions is made?
- Given that such a correction could be found, could we go back and make a correction to the gradient of the energy with respect to the support functions, with the knowledge that changing them would imply an automatic change in the density kernel?

In addition, one could ask some questions about the consequent changes to the Hamiltonian matrix:

- How much of the change in the Hamiltonian matrix, upon support function update, is due to a genuine change in the electronic structure, and how much is due to geometrical transformations (rotations within the same space)?
- If the latter part dominates, as it may be expected to in the latter stages of energy minimisation, then is it really necessary to perform an expensive re-evaluation of the Hamiltonian matrix each time the support functions or the density kernel change?
- When we are close to convergence, and changes to the spatial form of the support functions are expected to be small, would a computationally inexpensive geometrical update of the Hamiltonian matrix elements (or at least the density-independent components of those elements) suffice, rather than complete re-evaluation?

In this chapter, as well as exploring some interesting aspects of the non-orthogonal nature of the support functions, we will analytically investigate the consequences (for the density kernel and the Hamiltonian) of altering the support functions. We derive corrections to the density kernel and the Hamiltonian, on geometric grounds, which depend on the support function update vectors and which re-introduce the conservation of some desirable properties. We have implemented such corrections for the case of the density kernel, allowing us make some judgements on the first set of queries on quantitative grounds, and this is described in Chap. 6.

We refer the reader to Ref. [14] for a transparent exposition of the central concepts, definitions and expressions used in this chapter, and we freely acknowledge this latter work as the primary source of textbook differential geometry drawn upon. For an excellent introduction to the field of Riemannian surfaces, one which has aided our conceptual understanding of the meaning and significance of the covariant derivative and of Ricci's Lemma, and which motivated our introduction of the *support manifold*, we direct the reader to Ref. [15]. Further useful sources on differential geometry were Ref. [16] (whose notation we broadly adhere to), Ref. [17] (for affine connections and curvature tensors) and Ref. [18] (for a robust exposition on underlying concepts and whose useful bullet notation • which we employ to express the tensorial character of matrices in cases where explicit indices would complicate the exposition).

## 5.2 Tensor Calculus Applied to Electronic Structure Theory

The support functions, in order to describe the Kohn–Sham orbitals, lie in a Hilbert vector space which we term the *support space* and which has nontrivial properties which result from the nonorthogonality of the support functions. In this chapter, in order to describe such properties, we will be concerned with the geometry of a surface defined by the scalar function (generally complex valued, but we will quickly restrict ourselves to the real-valued case) of a $N$-tuplet of co-ordinate vectors,

$$|\mathcal{F}\rangle = \{\mathcal{F}\left(\mathbf{r}^1, \ldots, \mathbf{r}^N\right) \in \mathbb{C} \mid \mathbf{r}^\alpha \in \mathbb{R}^3 \mid \alpha \in \{1, \ldots, N\}\}. \tag{5.2}$$

The support space is equivalent to the cotangent space of the surface defined by $|\mathcal{F}\rangle$, that is

$$|\phi_\alpha\rangle = |\mathcal{F}^{,\alpha}\rangle = \frac{\partial|\mathcal{F}\rangle}{\partial\mathbf{r}^\alpha}, \tag{5.3}$$

or, making the spatial-dependence more explicit,

$$\phi_\alpha(\mathbf{r}) = \left.\frac{\partial\mathcal{F}\left(\mathbf{s}^1, \ldots, \mathbf{s}^\alpha, \ldots, \mathbf{s}^N\right)}{\partial\mathbf{s}^\alpha}\right|_{\left(\mathbf{s}^1, \ldots, \mathbf{s}^\alpha, \ldots, \mathbf{s}^N\right) = (\mathbf{r}, \ldots, \mathbf{r}, \ldots, \mathbf{r})}. \tag{5.4}$$

Here the index $\alpha$ is a label on position vectors in direct space and $N$ is the cardinality of support functions—assumed to be the number of distinct Kohn–Sham functions which need to be described. In practice, it is not necessary to compute the value of the function $\mathcal{F}$ in first-principles calculations, though it is useful to give some consideration to its properties, as we proceed to show. The surface defined by $\mathcal{F}$ is described here as the *support manifold* and it may be shown that the support manifold itself is a smooth surface in $\mathbb{R}^{3N} \otimes \mathbb{R}$ when the support functions are real-valued and linearly-independent [15]. Linear-independence of support functions (which does not imply orthogonality in general) is an important prerequisite for stable energy minimisation, perhaps fundamentally for this very reason, and we assume that that this desirable property is in place in what follows so that continuous derivatives of all orders may be computed.

The support functions form a set of covariant vectors as an immediate consequence of their definition in terms of $\mathcal{F}$, that is if we perform a transformation of the basis (in direct space) from $\mathbf{r}$ to $\tilde{\boldsymbol{r}}$, they transform with it according to

$$\phi_\alpha\left(\tilde{\boldsymbol{r}}\right) \equiv \frac{\partial\mathbf{r}}{\partial\tilde{\boldsymbol{r}}}\phi_\alpha(\mathbf{r}). \tag{5.5}$$

The support functions, naturally as elements of the cotangent space to the support manifold, behave under basis transformations in the same manner as the gradient vectors of a scalar function.

The space of support functions is accompanied by a dual space, the elements of which, *dual functions*, may be written in terms of the support functions as

$$|\phi_\alpha\rangle = |\phi^\beta\rangle S_{\beta\alpha} \quad \text{or} \quad |\phi^\alpha\rangle = |\phi_\beta\rangle S^{\beta\alpha}, \tag{5.6}$$

where $S_{\bullet\bullet}$ is a special matrix characteristic of the manifold known as the *metric tensor* and $S^{\bullet\bullet}$ is its inverse. The dual functions transform as contravariant vectors. This means that under the same transformation $\mathbf{r}$ to $\tilde{\boldsymbol{r}}$ they behave as the tangent vectors to the support manifold, that is

$$\phi^\alpha\left(\tilde{\boldsymbol{r}}\right) \equiv \frac{\partial \tilde{\boldsymbol{r}}}{\partial \mathbf{r}}\phi^\alpha\left(\mathbf{r}\right). \tag{5.7}$$

In practice, the frame in which the Hamiltonian is computed is conventionally defined as the covariant and the frame in which the elements of the matrix representation of the density-matrix, the density kernel, are defined is consequently the contravariant. Henceforth, a vector with a subscript denotes a covariant quantity and a superscript indicates contravariance. Pairs of the same index in the same expression, one subscript and one superscript, indicate that this index is summed over unless the index appears in parentheses.

Finally, to elucidate the meaning of the support manifold, suppose that it is everywhere smooth so that its cotangent and tangent spaces are well defined at each point on the surface. As a result, and since for each set of support functions we can find a global minimum of the total energy with respect to idempotent density kernels (using, for example, the LNVD technique [19–21]), there is a one-to-one mapping between each point on the surface and the best energy achievable (assuming non-degeneracy) with its tangent and cotangent vectors at that point. The Hohenberg–Kohn theorem [6] implies that if no further approximations are used (e.g., that the exact exchange-correlation functional were known, basis-set completeness can be achieved, the support functions and density kernel are not truncated etc.) and if we assume that the bijection between support manifold and energy is continuous (we suppose that it is, on grounds of physical intuition, but we cannot offer a proof) there is a minimum of the total energy with respect to points on the support manifold. The problem of extremising the total energy with respect to the support functions (and their corresponding optimal density kernels) becomes one of extremising $\mathcal{F}$ on the support manifold.

### 5.2.1 Tensorial Invariance

The support functions and their duals satisfy, by definition, the condition of biorthogonality

$$\phi_\alpha \bullet \phi^\beta = \delta_\alpha{}^\beta, \tag{5.8}$$

where $\bullet$ is the inner product between elements of the vector space of support functions and its corresponding vector space of dual functions. The natural inner product for quantum states is the Hilbert integral, so that the definition of support and dual functions in terms of each other becomes

$$\phi_\alpha \bullet \phi^\beta \equiv \langle\phi_\alpha|\phi^\beta\rangle = \int d\mathbf{r}\; \phi^*_\alpha(\mathbf{r})\, \phi^\beta(\mathbf{r}) = \delta_\alpha{}^\beta. \tag{5.9}$$

Combining Eqs. 5.6 and 5.9, we immediately find that the metric and its inverse are respectively given by the overlap matrix of support functions and duals, explicitly

$$S_{\alpha\beta} = \langle\phi_\alpha|\phi_\beta\rangle \quad \text{and} \quad S^{\alpha\beta} = \langle\phi^\alpha|\phi^\beta\rangle. \tag{5.10}$$

The identity operator on the Hilbert space spanned by the support functions (or duals) is expanded in terms of these functions by taking the product of the metric and its inverse, that is

$$\hat{1} = S_{\alpha\beta} S^{\beta\alpha} = \langle\phi_\alpha|\phi_\beta\rangle\langle\phi^\beta|\phi^\alpha\rangle = |\phi_\alpha\rangle\langle\phi^\alpha| = |\phi^\alpha\rangle\langle\phi_\alpha|. \tag{5.11}$$

Once defined, the metric, Eq. 5.10, together with its derivatives, fully describes the differential geometry of the support manifold. By definition, the support functions define the identity of the support manifold, and the covariant support functions and their contravariant duals are inter-related via the metric, so that

$$\begin{aligned} |\phi_\alpha\rangle S^{\alpha\beta}\langle\phi_\beta| &= |\phi_\alpha\rangle\langle\phi^\alpha|\phi^\beta\rangle\langle\phi_\beta| = \hat{1} \\ &= |\phi^\alpha\rangle\langle\phi_\alpha|\phi_\beta\rangle\langle\phi^\beta| = |\phi^\alpha\rangle S_{\alpha\beta}\langle\phi^\beta|. \end{aligned} \tag{5.12}$$

In order to illustrate what is meant by a tensorial invariant, we consider the trace of the product of the Hamiltonian $\hat{H}$ and the density-matrix, $\hat{\rho}$, which are both assumed to be completely described by the support functions. Using this definition of the identity operator, we can write the noninteracting energy (also known as the band-energy) as a tensor expression, as in

$$E = \mathbf{Tr}\left[\hat{H}\hat{\rho}\right] = |\phi^\alpha\rangle\langle\phi_\alpha|\hat{H}|\phi_\beta\rangle\langle\phi^\beta|\hat{\rho}|\phi^\gamma\rangle\langle\phi_\gamma| = H_{\alpha\beta} K^{\beta\alpha}. \tag{5.13}$$

The advantage of the tensorial expression of the noninteracting energy as a contraction of the covariant second-order Hamiltonian tensor $H_{\alpha\beta} = \langle\phi_\alpha|\hat{H}|\phi_\beta\rangle$ and the contravariant second-order tensor $K^{\beta\alpha} = \langle\phi^\beta|\hat{\rho}|\phi^\alpha\rangle$ is that the result is a scalar which is independent of the coordinate system. The scalar $E$ is then said to be a *tensorial invariant* and quantities such as energies, integrated occupancies and magnetisation densities, spectra, the Cartesian components of moments and forces etc. should possess this quality if they are to carry physical meaning (if they are to be independent, for example, under arbitrary linear-transformations among the support functions). This is very important to emphasise: physically meaningful quantities are only computed via pairwise contraction over indices with opposite tensor character.

Another scalar invariant of interest, that defined as

$$\delta s^2 = |\delta\phi^\alpha\rangle S_{\alpha\beta} \langle\delta\phi^\beta|, \tag{5.14}$$

is known as a *metric connection* on the Riemannian space (a space with a non-trivial overlap matrix $S_{\bullet\bullet}$) spanned by the vectors of support functions. The infinitesimal element $\delta s$ is interpreted as the tensorially invariant measure of distance between a function $\phi$ and the nearby function $\phi + \delta\phi$ [14].

A crucial point to note is that, as an immediate result of Eqs. 5.5 and 5.7, the derivative of an invariant quantity with respect to a dual function is an element of the space of support functions. For example, noting the independence of real and imaginary parts of the support functions, we have

$$\begin{aligned} E_{,|\gamma\rangle} &\equiv \frac{\partial E}{\partial|\phi^\gamma\rangle} = \frac{\partial H_{\alpha\beta} K^{\beta\alpha}}{\partial|\phi^\gamma\rangle} = H_{\alpha\beta} \langle\phi^\beta|\hat{\rho}\delta_\gamma{}^\alpha \\ &= H_{\gamma\beta} \langle\phi^\beta|\hat{\rho}|\phi^\delta\rangle\langle\phi_\delta| = \langle\phi_\delta| H_{\gamma\beta} K^{\beta\delta}. \end{aligned} \tag{5.15}$$

The converse is also true, that is that the derivative of an invariant quantity with respect to a support function yields an element of the space of dual functions, a contravariant vector as we might expect, for example as in

$$\begin{aligned} E^{,|\gamma\rangle} &\equiv \frac{\partial E}{\partial|\phi_\gamma\rangle} = \frac{\partial H_{\alpha\beta} K^{\beta\alpha}}{\partial|\phi_\gamma\rangle} = \langle\phi_\alpha|\hat{H}\delta^\gamma{}_\beta K^{\beta\alpha} \\ &= \langle\phi_\alpha|\hat{H}|\phi_\delta\rangle\langle\phi^\delta|K^{\gamma\alpha} = \langle\phi^\delta| H_{\alpha\delta} K^{\gamma\alpha}. \end{aligned} \tag{5.16}$$

## 5.3 Partial Differentiation of Tensors

The partial derivative of a covariant or contravariant vector (that is with respect to a representation vector—the co-ordinate basis $\mathbf{r}$ is hereafter only used to define the inner product), or of a higher order fully covariant, fully contravariant or mixed tensor, does not itself provide a tensor, in general [14]. The components of the resulting expressions, which are known as the components of coordinate charts, do not transform to the corresponding components of the derivative of the transformed object (i.e., basis transformation and partial differentiation do not commute in general).

The mathematical object which connects tangent spaces infinitesimally near to each other, restoring this commutativity, is known as an *affine connection*. Geometrically, we can understand this as a mechanism for using the curved support manifold as a fixed surface along which one tangent plane is rolled to another (infinitesimally proximate tangent plane), and so with respect to which the derivative can be unambiguously computed. These affine connections add correction terms to the partial derivatives, so that they become *absolute derivatives* and produce objects whose components transform covariantly or contravariantly, as the case may be. We henceforth employ the commonly-used shorthand notation $,\bullet$ for partial derivation and $;\bullet$

for absolute differentiation. The latter will be defined below for the special case of the *Levi-Civita connection*.

For the sake of simplicity, we hereafter restrict ourselves to the real-valued support functions $\{\phi_{\bullet}\}$ widely used in linear-scaling ab initio methods. The results here do not necessarily all apply straightforwardly to the case of complex support functions, particularly with regards to the torsion of the manifold. For the case of real support functions, the Hermitian metric tensor $S_{\alpha\beta}$ becomes symmetric and the distance measure is then uniquely defined. Thus, we may abandon the distinction between support functions and their complex conjugates, as well as dual functions and their respective complex conjugates, so that the Dirac "braket" notation is henceforth suppressed except where it conveniently expresses an inner product.

The support functions and duals are not independent (the tangent and cotangent spaces are unambiguously defined with respect to each other via the metric tensor) and thus one has non-zero partial derivatives with respect to the other. The partial derivative of the metric with respect to covariant support functions is given by

$$S_{\alpha\beta}^{,\gamma} = \phi_\alpha \delta_\beta^\gamma + \delta_\alpha^\gamma \phi_\beta \tag{5.17}$$

and that of the inverse metric with respect to contravatiant dual functions is

$$S^{\alpha\beta}_{,\gamma} = \phi^\alpha \delta_\gamma^\beta + \delta_\gamma^\alpha \phi^\beta . \tag{5.18}$$

Technically, the relationship between support functions and their duals is an example of a holonomic constraint, that is one for which the Lie commutator

$$\begin{aligned}
\gamma_{\beta\gamma}^{\alpha} &= [\phi_\beta, \phi_\gamma]\,\phi^\alpha \\
&\equiv \left[\left(\phi^\beta \frac{\partial}{\partial\phi^\beta}\right)\left(\phi^\gamma \frac{\partial}{\partial\phi^\gamma}\right) - \left(\phi^\gamma \frac{\partial}{\partial\phi^\gamma}\right)\left(\phi^\beta \frac{\partial}{\partial\phi^\beta}\right)\right]\phi^\alpha \\
&= \left[\left(\phi^\beta \frac{\partial}{\partial\phi^\beta}\right)\left(\phi^\gamma \delta_\gamma^\alpha\right) - \left(\phi^\gamma \frac{\partial}{\partial\phi^\gamma}\right)\left(\phi^\beta \delta_\beta^\alpha\right)\right] \\
&= \left[\left(\phi^\beta \delta_\beta^\gamma \delta_\gamma^\alpha\right) - \left(\phi^\gamma \delta_\gamma^\beta \delta_\beta^\alpha\right)\right] = \phi^\alpha - \phi^\alpha = 0
\end{aligned} \tag{5.19}$$

vanishes and in which case the evaluation of tensorial derivatives simplifies considerably.

Taking the partial derivative of the identity operator, Eq. 5.12, with respect to the support functions (top line) or dual functions (bottom line), while making the necessary assumption that the identity operator of the support manifold itself does not change to first order in the support functions, we find that

$$\begin{aligned}
&\delta_\alpha^\gamma S^{\alpha\beta}\phi_\beta + \phi_\alpha S^{\alpha\beta,\gamma}\phi_\beta + \phi_\alpha S^{\alpha\beta}\delta_\beta^\gamma = 0 = \frac{\partial \hat{I}}{\partial \phi_\gamma} \quad \text{and} \\
&\delta_\gamma^\alpha S_{\alpha\beta}\phi^\beta + \phi^\alpha S_{\alpha\beta,\gamma}\phi^\beta + \phi^\alpha S_{\alpha\beta}\delta_\gamma^\beta = 0 = \frac{\partial \hat{I}}{\partial \phi^\gamma} .
\end{aligned} \tag{5.20}$$

Consequently, we may conclude that, in the special case of a real complete set of support functions, the fully contravariant and covariant (in a manner of speaking—these do not transform as tensors) first partial derivatives of the metric are given, respectively, by

$$S^{\alpha\beta,\gamma} = -\phi^{\alpha} S^{\beta\gamma} - \phi^{\beta} S^{\alpha\gamma} \quad \text{and} \tag{5.21}$$

$$S_{\alpha\beta,\gamma} = -\phi_{\alpha} S_{\beta\gamma} - \phi_{\beta} S_{\alpha\gamma}. \tag{5.22}$$

The opposite signs appearing in the derivative of the metric with respect to its spanning functions, Eqs 5.17 and 5.18, and with respect to the set conjugate to its spanning functions, Eqs. 5.21 and 5.22, are an immediate example of the aforementioned non-tensorial character of partial derivatives: partial differentiation operators do not simply commute with changes of index position using the metric tensor.

The assertion that the identity operator may be considered constant while computing the derivatives of the metric is, undeniably, a somewhat bold one. We do not expect this to hold rigorously in the early stages of a total-energy minimisation procedure, though it is bound to attain validity as we approach convergence. As we go on to show, however, it is analytically and numerically favourable to assume first-order preservation of the support space, and make use of the results, rather than the alternative of neglecting the geometric ramifications of support function update entirely.

## 5.4 A Metric Connection on the Support Manifold

Our next task is to check whether the support manifold admits a metric connection, to wit, whether it admits an affine connection which preserves the metric tensor under first derivatives. If this is the case then the metric is the special tensor preserved by absolute differentiation with respect to the tangent or cotangent vectors of the support manifold at all points—the measure of distance is preserved under differentiation.

In order to determine if this desirable property holds, we consider the Christoffel symbols (of the first kind), which are defined as

$$\begin{aligned}\Gamma_{\alpha\beta\gamma} &\equiv \frac{1}{2}\left(S_{\alpha\beta,\gamma} + S_{\alpha\gamma,\beta} - S_{\beta\gamma,\alpha}\right) \\ &= \frac{1}{2}\left(-\phi_{\alpha} S_{\beta\gamma} - \phi_{\beta} S_{\alpha\gamma} - \phi_{\alpha} S_{\gamma\beta} - \phi_{\gamma} S_{\alpha\beta} + \phi_{\beta} S_{\gamma\alpha} + \phi_{\gamma} S_{\beta\alpha}\right) \\ &= -\phi_{\alpha} S_{\beta\gamma}.\end{aligned} \tag{5.23}$$

Here we freely use the summation convention for superscripts and subscripts. We may define another useful set of Christoffel symbols (those known as the second kind) by

$$\Gamma^{\gamma}_{\alpha\beta} = S^{\gamma\delta}\Gamma_{\delta\alpha\beta} = -\phi^{\gamma} S_{\alpha\beta} \tag{5.24}$$

in spite of the fact that the Christoffel symbols do not generally transform as tensors. We have made use of the symmetry of the metric tensor in the above (we could not do so in the case of complex-valued support functions, however), and we do so again in satisfying ourselves that the torsion tensor defined as

$$T^{\gamma}_{\alpha\beta} \equiv \Gamma^{\gamma}_{\alpha\beta} - \Gamma^{\gamma}_{\beta\alpha} - \gamma^{\gamma}_{\alpha\beta} = S_{\alpha\beta} - S_{\beta\alpha} - 0 \tag{5.25}$$

is zero on the whole surface. We may therefore conclude that the support manifold is torsion-free in the case of real-valued sets of support functions cotangent to support manifolds admitting completeness preservation. We reiterate that in the case of complex support functions the metric is Hermitian but not generally symmetric and so the Christoffel symbol is not as simple as in the real-valued case.

The Christoffel symbols serve as corrections to the partial derivatives which provide derivatives, known as *absolute derivatives*, whose components transform as the components of either covariant or contravariant vectors. The action of absolute derivatives on tensors of general rank and tensor character, where we again remind the reader of the shorthand notation ,• for partial derivation and ;• for absolute differentiation, may be generalised from a well-known result for the covariant first derivative on the mixed rank-2 tensor, given by

$$A^{\beta}_{\alpha;\gamma} = A^{\beta}_{\alpha,\gamma} + \Gamma^{\beta}_{\gamma\delta} A^{\delta}_{\alpha} - \Gamma^{\delta}_{\alpha\gamma} A^{\beta}_{\delta}, \tag{5.26}$$

and the contravariant first derivate defined in terms of it, namely

$$A^{\beta;\gamma}_{\alpha} = S^{\gamma\delta} A^{\beta}_{\alpha;\delta} = S^{\gamma\delta} A^{\beta}_{\alpha,\delta} + S^{\gamma\delta} \Gamma^{\beta}_{\delta\epsilon} A^{\epsilon}_{\alpha} - S^{\gamma\delta} \Gamma^{\epsilon}_{\alpha\delta} A^{\beta}_{\epsilon}. \tag{5.27}$$

In this way, we may evaluate derivatives on surfaces in the space of support functions which account for the curvature of the manifold and which thus transform as tensors. The terms due to partial derivates take into account the change in tensors with respect to the local tangent (or cotangent—depending on the index positions) space, while the terms due to the Christoffel symbols, in effect, correct for the infinitesimal rocking of the tangent (and consequently cotangent) space on the support manifold. As a result, tensorial integrity is preserved under absolute differentiation by ensuring that corrections are made as we move along the support manifold in order to ensure that we always remain tensorially consistent with the local tangent and cotangent spaces.

Returning to the evaluation of the absolute derivative of the metric tensor itself, using Eq. 5.26, we reproduce the well-known result (Ricci's Lemma) that the terms due to the partial derivative and due to the Christoffel symbols cancel, so that

$$\begin{aligned} S_{\alpha\beta;\gamma} &= S_{\alpha\beta,\gamma} - \Gamma^{\delta}_{\alpha\gamma} S_{\delta\beta} - \Gamma^{\delta}_{\beta\gamma} S_{\delta\alpha} \\ &= -\phi_{\alpha} S_{\beta\gamma} - \phi_{\beta} S_{\alpha\gamma} + \phi_{\alpha} \delta^{\delta}_{\gamma} S_{\delta\beta} + \phi_{\beta} \delta^{\delta}_{\gamma} S_{\delta\alpha} \\ &= -\phi_{\alpha} S_{\beta\gamma} - \phi_{\beta} S_{\alpha\gamma} + \phi_{\alpha} S_{\gamma\beta} + \phi_{\beta} S_{\gamma\alpha} \\ &= 0, \end{aligned} \tag{5.28}$$

as required for a metric connection. This provides a *post hoc* justification for our choice of Christoffel symbol, namely that known as the Levi-Civita connection, since it is the unique symbol which leaves the metric invariant under covariant differentiation (that is absolute differentiation with respect to a contravariant vector) on a torsion-free metric-compatible Riemannian manifold [14]. Geometrically speaking, then, the distance measure on the manifold of complete sets of support functions does not change to first order with the support functions when the necessary geometric corrections are used.

## 5.5 Variation of the Density Kernel and Hamiltonian

Let us now consider how two important tensors in linear-scaling density functional theory transform with changes in the support functions, namely the covariant Hamiltonian matrix and the contravariant density kernel matrix. These are given by the expressions, respectively,

$$H_{\alpha\beta} = \langle \phi_\alpha | \hat{H} | \phi_\beta \rangle \quad \text{and} \quad K^{\alpha\beta} = \langle \phi^\alpha | \hat{\rho} | \phi^\beta \rangle, \tag{5.29}$$

where $\hat{H}$ is the Hamiltonian operator, $\hat{\rho}$ is the Kohn–Sham density-matrix and the spin and **k**-point indices are suppressed for brevity.

### 5.5.1 *Uncorrected Matrix Updates*

One may directly compute the first partial covariant and first partial contravariant derivatives of these quantities with respect to (real-valued) support functions. In the case of the Hamiltonian matrix we find, for the contravariant derivative, that

$$\begin{aligned} H_{\alpha\beta}^{;\gamma} &= \frac{\partial}{\partial \phi_\gamma} \left[ \langle \phi_\alpha | \hat{H} | \phi_\beta \rangle \right] \\ &= \hat{H} \phi_\beta \delta_\alpha^\gamma + \hat{H} \phi_\alpha \delta_\beta^\gamma = \phi^\epsilon \left( H_{\epsilon\beta} \delta_\alpha^\gamma + H_{\epsilon\alpha} \delta_\beta^\gamma \right). \end{aligned} \tag{5.30}$$

For the covariant derivative, on the other hand, noting that the tangent and cotangent spaces explicitly interdepend via the metric tensor, we obtain

$$\begin{aligned} H_{\alpha\beta,\gamma} &= \frac{\partial}{\partial \phi^\gamma} \left[ \langle \phi_\alpha | \hat{H} | \phi_\beta \rangle \right] = \frac{\partial}{\partial \phi^\gamma} \left[ S_{\alpha\delta} \langle \phi^\delta | \hat{H} | \phi^\epsilon \rangle S_{\epsilon\beta} \right] \\ &= S_{\alpha\delta,\gamma} H^{\delta\epsilon} S_{\epsilon\beta} + S_{\alpha\delta} \delta_\gamma^\delta \hat{H} \phi^\epsilon S_{\epsilon\beta} + S_{\alpha\delta} \phi^\delta \hat{H} \delta_\gamma^\epsilon S_{\epsilon\beta} + S_{\alpha\delta} H^{\delta\epsilon} S_{\epsilon\beta,\gamma} \\ &= \left( -\phi_\alpha S_{\delta\gamma} - \phi_\delta S_{\alpha\gamma} \right) H^{\delta\epsilon} S_{\epsilon\beta} + S_{\alpha\gamma} \hat{H} \phi_\beta + S_{\gamma\beta} \hat{H} \phi_\alpha \\ &\quad + S_{\alpha\delta} H^{\delta\epsilon} \left( -\phi_\epsilon S_{\beta\gamma} - \phi_\beta S_{\epsilon\gamma} \right) \quad \text{(having used Eq. 5.22)} \end{aligned}$$

$$
\begin{aligned}
&= -\phi_\alpha H_{\gamma\beta} - S_{\alpha\gamma}\hat{H}\phi_\beta + S_{\alpha\gamma}\hat{H}\phi_\beta + S_{\beta\gamma}\hat{H}\phi_\alpha \\
&\quad - \phi_\beta H_{\alpha\gamma} - S_{\beta\gamma}\hat{H}\phi_\alpha = -\phi_\alpha H_{\beta\gamma} - \phi_\beta H_{\alpha\gamma}.
\end{aligned} \tag{5.31}
$$

For the density kernel, we find that the expressions take a similar form, with the index positions inter-related by index symmetries. The contravariant derivative of the density kernel is given by

$$
\begin{aligned}
K^{\alpha\beta,\gamma} &= \frac{\partial}{\partial\phi_\gamma}\left[\langle\phi^\alpha|\hat{\rho}|\phi^\beta\rangle\right] = \frac{\partial}{\partial\phi_\gamma}\left[S^{\alpha\delta}\langle\phi_\delta|\hat{\rho}|\phi_\epsilon\rangle S^{\epsilon\beta}\right] \\
&= S^{\alpha\delta,\gamma} K_{\delta\epsilon} S^{\epsilon\beta} + S^{\alpha\delta}\delta_\delta^\gamma\hat{\rho}\phi_\epsilon S^{\epsilon\beta} + S^{\alpha\delta}\phi_\delta\hat{\rho}\delta_\epsilon^\gamma S^{\epsilon\beta} + S^{\alpha\delta} K_{\delta\epsilon} S^{\epsilon\beta,\gamma} \\
&= \left(-\phi^\alpha S^{\delta\gamma} - \phi^\delta S^{\alpha\gamma}\right) K_{\delta\epsilon} S^{\epsilon\beta} + S^{\alpha\gamma}\hat{\rho}\phi^\beta + S^{\gamma\beta}\hat{\rho}\phi^\alpha \\
&\quad + S^{\alpha\delta} K_{\delta\epsilon}\left(-\phi^\epsilon S^{\beta\gamma} - \phi^\beta S^{\epsilon\gamma}\right) \quad \text{(having used Eq. 5.21)} \\
&= -\phi^\alpha K^{\gamma\beta} - S^{\alpha\gamma}\hat{\rho}\phi^\beta + S^{\alpha\gamma}\hat{\rho}\phi^\beta + S^{\beta\gamma}\hat{\rho}\phi^\alpha \\
&\quad - \phi^\beta K^{\alpha\gamma} - S^{\beta\gamma}\hat{\rho}\phi^\alpha \\
&= -\phi^\alpha K^{\beta\gamma} - \phi^\beta K^{\alpha\gamma}
\end{aligned} \tag{5.32}
$$

and the corresponding covariant derivative is provided by

$$
\begin{aligned}
K^{\alpha\beta}_{,\gamma} &= \frac{\partial}{\partial\phi^\gamma}\left[\langle\phi^\alpha|\hat{K}|\phi^\beta\rangle\right] \\
&= \hat{\rho}\phi^\beta\delta_\gamma^\alpha + \hat{\rho}\phi_\alpha\delta_\gamma^\beta = \phi_\epsilon\left(K^{\epsilon\beta}\delta_\gamma^\alpha + K^{\epsilon\alpha}\delta_\gamma^\beta\right).
\end{aligned} \tag{5.33}
$$

The first order change in the Hamiltonian and the density kernel upon changes $\Delta\phi_\bullet$ of the support functions, considering only partial differentials ($\tilde{\Delta}$), and assuming that the operators themselves are unchanged by the update in support functions (this holds exactly if the support functions merely undergo a unitary transformation) may now be computed. These changes to the Hamiltonian and density-kernel are furnished, respectively, by

$$
\begin{aligned}
\tilde{\Delta} H_{\alpha\beta} &= \langle\Delta\phi_\gamma|H^{;\gamma}_{\alpha\beta}\rangle = \langle\Delta\phi_\gamma|\phi^\epsilon\rangle\left(H_{\epsilon\beta}\delta_\alpha^\gamma + H_{\epsilon\alpha}\delta_\beta^\gamma\right) \\
&= \langle\Delta\phi_\alpha|\phi^\epsilon\rangle H_{\epsilon\beta} + H_{\alpha\epsilon}\langle\phi^\epsilon|\Delta\phi_\beta\rangle
\end{aligned} \tag{5.34}
$$

and

$$
\begin{aligned}
\tilde{\Delta} K^{\alpha\beta} &= \langle\Delta\phi_\gamma|K^{\alpha\beta,\gamma}\rangle = \langle\Delta\phi_\gamma|\phi_\epsilon\rangle\left(-S^{\epsilon\alpha}K^{\beta\gamma} - S^{\epsilon\beta}K^{\alpha\gamma}\right) \\
&= -S^{\alpha\epsilon}\langle\phi_\epsilon|\Delta\phi_\gamma\rangle K^{\gamma\beta} - K^{\alpha\gamma}\langle\Delta\phi_\gamma|\phi_\epsilon\rangle S^{\epsilon\beta}.
\end{aligned} \tag{5.35}
$$

The matrices $\left(H + \tilde{\Delta} H\right)_{\bullet\bullet}$ and $\left(K + \tilde{\Delta} K\right)^{\bullet\bullet}$, unfortunately, are not generally those that we require, since they are not always guaranteed to correspond to the representations of the Hamiltonian and density-matrix in a set of support functions (dual functions) which are cotangent (tangent) to any point on the support manifold.

In general, the matrices will have departed from the tangent and cotangent frames at points on the support manifold infinitesimally close to that at which the partial derivatives are computed, instead of being consistently transported along a geodesic line connecting the points. As a result, they may not always act as tensors with respect to the updated functions $(\phi + \Delta\phi)_{\bullet}$.

This point is important to emphasise: if tensors in the frame of support functions (or their duals) are updated with uncorrected partial derivatives in order to account for changes in the support functions or simply not corrected at all, the resulting matrices will no longer, for general support manifolds, behave as tensors at all. In order to correct the matrices so that they are guaranteed to transform as tensors for all geometries, and so that they may be contracted properly with the support functions, we must again turn to the Levi-Civita connection, which provides the requisite expressions for absolute covariant and contravariant differentiation, i.e., which preserves the tensorial character of matrices by construction.

### 5.5.2 Geometrically Corrected Matrix Updates

Applying covariant absolute differentiation to the Hamiltonian $H_{\bullet\bullet}$ and density kernel $K^{\bullet\bullet}$, using the standard expressions for covariant and contravariant rank-two tensors, respectively, we find that

$$H_{\alpha\beta;\gamma} = H_{\alpha\beta,\gamma} - H_{\epsilon\beta}\Gamma^{\epsilon}_{\alpha\gamma} - H_{\alpha\epsilon}\Gamma^{\epsilon}_{\beta\gamma} \quad \text{and}$$

$$K^{\alpha\beta}_{;\gamma} = y K^{\alpha\beta}_{,\gamma} + K^{\epsilon\beta}\Gamma^{\alpha}_{\epsilon\gamma} + K^{\alpha\epsilon}\Gamma^{\beta}_{\epsilon\gamma}. \tag{5.36}$$

A previously mentioned and computationally useful result, known as *Ricci's Lemma*, is that since the metric tensor is invariant under absolute differentiation with respect to local tangent vectors, it freely commutes with absolute differentiation. Thus, we can generally write the covariant derivatives above in terms of contravariant derivatives simply as

$$H_{\alpha\beta}^{;\gamma} = H_{\alpha\beta;\delta}S^{\delta\gamma} = H_{\alpha\beta,\delta}S^{\delta\gamma} - H_{\epsilon\beta}\Gamma^{\epsilon}_{\alpha\delta}S^{\delta\gamma} - H_{\alpha\epsilon}\Gamma^{\epsilon}_{\beta\delta}S^{\delta\gamma}, \tag{5.37}$$

$$K^{\alpha\beta;\gamma} = K^{\alpha\beta}_{;\delta}S^{\delta\gamma} = K^{\alpha\beta}_{,\delta}S^{\delta\gamma} + K^{\epsilon\beta}\Gamma^{\alpha}_{\epsilon\delta}S^{\delta\gamma} + K^{\alpha\epsilon}\Gamma^{\beta}_{\epsilon\delta}S^{\delta\gamma}. \tag{5.38}$$

We note that it is *not* implied by the above that the individual elements making up the absolutely differentiated tensor must individually commute with the metric. In particular, it is generally the case that

$$H_{\alpha\beta}^{;\gamma} \neq H_{\alpha\beta,\delta}S^{\delta\gamma} \quad \text{and} \quad K^{\alpha\beta,\gamma} \neq K^{\alpha\beta}_{,\delta}S^{\delta\gamma}, \tag{5.39}$$

since the metric does not generally commute with partial derivatives. Rather, we may readily assure ourselves that Ricci's Lemma holds for the sum of the contributions to the absolute derivatives, as required, by checking that the sum of the commutators between the metric and the each of the individual terms in the derivative vanishes.

The strategy, then, for computing contravariant absolute derivatives (required since we usually deal explicitly with the support functions, and not their duals) is to compute the covariant derivative in full, and then raise the absolute differentiation index, only afterwards, using the contravariant metric tensor.

The required expression for the absolute derivative of the Hamiltonian thereby reduces to

$$\begin{aligned} H_{\alpha\beta}^{;\gamma} =& -\phi_\alpha H_{\beta\delta} S^{\delta\gamma} - \phi_\beta H_{\alpha\delta} S^{\delta\gamma} \\ & - H_{\epsilon\beta}\left(-\phi_\epsilon S_{\alpha\delta}\right) S^{\delta\gamma} - H_{\alpha\epsilon}\left(-\phi_\epsilon S_{\delta\beta}\right) S^{\delta\gamma} \\ =& -\phi_\alpha H_{\beta\delta} S^{\delta\gamma} - \phi_\beta H_{\alpha\delta} S^{\delta\gamma} + \phi^\epsilon \left(H_{\epsilon\beta} S_{\alpha\delta} S^{\delta\gamma} + H_{\alpha\epsilon} S_{\delta\beta} S^{\delta\gamma}\right) \\ =& -\phi_\alpha H_{\beta\delta} S^{\delta\gamma} - \phi_\beta H_{\alpha\delta} S^{\delta\gamma} + \phi^\epsilon \left(H_{\epsilon\beta} \delta_\alpha^\gamma + H_{\epsilon\alpha} \delta_\beta^\gamma\right), \end{aligned} \tag{5.40}$$

while that for the density kernel collapses to

$$\begin{aligned} K^{\alpha\beta;\gamma} =& \phi_\epsilon \left(K^{\epsilon\beta} \delta_\delta^\alpha + K^{\epsilon\alpha} \delta_\delta^\beta\right) S^{\delta\gamma} \\ & + K^{\epsilon\beta}\left(-\phi^\alpha S_{\epsilon\delta}\right) S^{\delta\gamma} + K^{\alpha\epsilon}\left(-\phi^\beta S_{\epsilon\delta}\right) S^{\delta\gamma} \\ =& \phi_\epsilon \left(K^{\epsilon\beta} S^{\alpha\gamma} + K^{\epsilon\alpha} S^{\beta\gamma}\right) - \phi^\alpha K^{\beta\epsilon} \delta_\epsilon^\gamma - \phi^\beta K^{\alpha\epsilon} \delta_\epsilon^\gamma \\ =& \phi_\epsilon \left(K^{\epsilon\beta} S^{\alpha\gamma} + K^{\epsilon\alpha} S^{\beta\gamma}\right) - \phi^\alpha K^{\beta\gamma} - \phi^\beta K^{\alpha\gamma}. \end{aligned} \tag{5.41}$$

Returning next to our expressions for the first order change in the Hamiltonian upon support function update, we may compute the change matrix which respects the curved nature of the support manifold. This differs to that computed when only the change along the tangent plane is considered, that is the previously computed and erroneous $\tilde{\Delta} H_{\bullet\bullet}$, and, in fact, we find that

$$\begin{aligned} \Delta H_{\alpha\beta} =& \langle \Delta\phi_\gamma | H_{\alpha\beta}^{;\gamma} \rangle \\ =& y\langle \Delta\phi_\gamma | \phi^\epsilon \rangle \left(H_{\epsilon\beta}\delta_\alpha^\gamma + H_{\epsilon\alpha}\delta_\beta^\gamma\right) \\ & - \langle \Delta\phi_\gamma | \phi_\alpha \rangle H_{\beta\delta} S^{\delta\gamma} - \langle \Delta\phi_\gamma | \phi_\beta \rangle H_{\alpha\delta} S^{\delta\gamma} \\ =& \langle \Delta\phi_\alpha | \phi_\zeta \rangle S^{\zeta\epsilon} H_{\epsilon\beta} + H_{\alpha\epsilon} S^{\epsilon\zeta} \langle \phi_\zeta | \Delta\phi_\beta \rangle \\ & - \langle \phi_\alpha | \Delta\phi_\gamma \rangle S^{\gamma\delta} H_{\delta\beta} - H_{\alpha\delta} S^{\delta\gamma} \langle \Delta\phi_\gamma | \phi_\beta \rangle \\ =& H_{\alpha\delta} S^{\delta\gamma} \left(\langle \phi_\gamma | \Delta\phi_\beta \rangle - \langle \Delta\phi_\gamma | \phi_\beta \rangle\right) \\ & + \left(\langle \Delta\phi_\alpha | \phi_\gamma \rangle - \langle \phi_\alpha | \Delta\phi_\gamma \rangle\right) S^{\gamma\delta} H_{\delta\beta} \\ \Rightarrow \Delta H_{\alpha\beta} =& -2 H_{\alpha\delta} S^{\delta\gamma} \langle \Delta\phi | \phi \rangle_{[\gamma\beta]} + 2\langle \Delta\phi | \phi \rangle_{[\alpha\gamma]} S^{\gamma\delta} H_{\delta\beta}. \end{aligned} \tag{5.42}$$

Here, the antisymmetrisation of the support function change vector and the support function vector at which the derivatives are evaluated is given by the shorthand

$$\langle \Delta\phi|\phi\rangle_{[\gamma\epsilon]} = \frac{1}{2}\left(\langle\Delta\phi_\gamma|\phi_\epsilon\rangle - \langle\phi_\gamma|\Delta\phi_\epsilon\rangle\right). \tag{5.43}$$

In the case of the total change of the density kernel, we find that the required expressions differ only slightly, insofar as

$$\begin{aligned}
\Delta K^{\alpha\beta} &= \langle\Delta\phi_\gamma|K^{\alpha\beta;\gamma}\rangle \\
&= \langle\Delta\phi_\gamma|\phi_\epsilon\rangle\left(K^{\epsilon\beta}S^{\alpha\gamma} + K^{\epsilon\alpha}S^{\beta\gamma}\right) \\
&\quad - \langle\Delta\phi_\gamma|\phi^\alpha\rangle K^{\beta\epsilon}\delta^\gamma_\epsilon - \langle\Delta\phi_\gamma|\phi^\beta\rangle K^{\alpha\epsilon}\delta^\gamma_\epsilon \\
&= S^{\alpha\gamma}\langle\Delta\phi_\gamma|\phi_\epsilon\rangle K^{\epsilon\beta} + K^{\alpha\epsilon}\langle\phi_\epsilon|\Delta\phi_\gamma\rangle S^{\gamma\beta} \\
&\quad - \langle\phi^\alpha|\Delta\phi_\epsilon\rangle K^{\epsilon\beta} - K^{\alpha\epsilon}\langle\Delta\phi_\epsilon|\phi^\beta\rangle \\
&= S^{\alpha\gamma}\left(\langle\Delta\phi_\gamma|\phi_\epsilon\rangle - \langle\phi_\gamma|\Delta\phi_\epsilon\rangle\right)K^{\epsilon\beta} \\
&\quad + K^{\alpha\epsilon}\left(\langle\phi_\epsilon|\Delta\phi_\gamma\rangle - \langle\Delta\phi_\epsilon|\phi_\gamma\rangle\right)S^{\gamma\beta} \\
\Rightarrow \Delta K^{\alpha\beta} &= 2S^{\alpha\gamma}\langle\Delta\phi|\phi\rangle_{[\gamma\epsilon]}K^{\epsilon\beta} - 2K^{\alpha\epsilon}\langle\Delta\phi|\Delta\phi\rangle_{[\epsilon\gamma]}S^{\gamma\beta}.
\end{aligned} \tag{5.44}$$

We will return to these corrections in a numerical study in which we investigate their effect on the convergence behaviour of the density functional theory solver known as ONETEP, in which the support functions for the Kohn–Sham orbitals, and the corresponding density kernel, are individually optimised in order to minimise the total electronic energy.

In this study, due to a subtlety of the energy minimisation procedure, it will prove to be useful to individually investigate the effect of terms which make up this correction. Namely, we will consider the update term due to the tensorially corrected contravariant partial derivatives,

$$\begin{aligned}
\Delta^{\mathcal{P}} H_{\alpha\beta} &= \langle\Delta\phi_\gamma|H_{\alpha\beta,\delta}\rangle S^{\delta\gamma} \\
&= -H_{\alpha\delta}S^{\delta\gamma}\langle\Delta\phi_\gamma|\phi_\beta\rangle - \langle\phi_\alpha|\Delta\phi_\gamma\rangle S^{\gamma\delta}H_{\delta\beta},
\end{aligned} \tag{5.45}$$

$$\begin{aligned}
\Delta^{\mathcal{P}} K^{\alpha\beta} &= \langle\Delta\phi_\gamma|K^{\alpha\beta}_{,\delta}\rangle S^{\delta\gamma} \\
&= S^{\alpha\gamma}\langle\Delta\phi_\gamma|\phi_\epsilon\rangle K^{\epsilon\beta} + K^{\alpha\epsilon}\langle\phi_\epsilon|\Delta\phi_\gamma\rangle S^{\gamma\beta},
\end{aligned} \tag{5.46}$$

separately to a term due to the Christoffel symbols, the latter being provided by

$$\begin{aligned}
\Delta^{\mathcal{C}} H_{\alpha\beta} &= -\langle\Delta\phi_\gamma|H_{\epsilon\beta}\Gamma^\epsilon_{\alpha\delta}\rangle S^{\delta\gamma} - \langle\Delta\phi_\gamma|H_{\alpha\epsilon}\Gamma^\epsilon_{\beta\delta}\rangle S^{\delta\gamma} \\
&= H_{\alpha\delta}S^{\delta\gamma}\langle\phi_\gamma|\Delta\phi_\beta\rangle + \langle\Delta\phi_\alpha|\phi_\gamma\rangle S^{\gamma\delta}H_{\delta\beta},
\end{aligned} \tag{5.47}$$

$$\begin{aligned}
\Delta^{\mathcal{C}} K^{\alpha\beta} &= \langle\Delta\phi_\gamma|K^{\epsilon\beta}\Gamma^\alpha_{\epsilon\delta}\rangle S^{\delta\gamma} + \langle\Delta\phi_\gamma|K^{\alpha\epsilon}\Gamma^\beta_{\epsilon\delta}\rangle S^{\delta\gamma} \\
&= -S^{\alpha\gamma}\langle\phi_\gamma|\Delta\phi_\epsilon\rangle K^{\epsilon\beta} - K^{\alpha\epsilon}\langle\Delta\phi_\epsilon|\phi_\gamma\rangle S^{\gamma\beta},
\end{aligned} \tag{5.48}$$

where together these sum to

$$\Delta H_{\alpha\beta} = \Delta^{\mathcal{P}} H_{\alpha\beta} + \Delta^{\mathcal{C}} H_{\alpha\beta} \quad \text{and}$$
$$\Delta K^{\alpha\beta} = \Delta^{\mathcal{P}} K^{\alpha\beta} + \Delta^{\mathcal{C}} K^{\alpha\beta}. \tag{5.49}$$

A geometric interpretation of each component, in the case of the density kernel update, is graphically illustrated in Fig. 5.1.

The former correction, $\Delta^{\mathcal{P}}$, is responsible for changes in the density kernel due to transformations of the dual functions in their local tangent space for a fixed density-matrix operator. Conventionally, such changes are neglected and so it is assumed that the density kernel itself remains unchanged during support function update and not the density-matrix operator. Of course, it is never possible to completely preserve the density-matrix explicitly when the space spanned by the support functions is allowed to change, however this term allows us to remove any spurious change in the density matrix due to unitary transformations among the dual functions.

The latter correction, $\Delta^{\mathcal{C}}$, on the other hand, corrects for the curved nature of the support manifold by returning the density kernel to the tangent space of the manifold at its updated point of contact. Without this term, the density kernel would no longer act as a tensor, and contracting it with the updated support functions would result in corruption of the total occupancy and idempotency of the density-matrix (this is evident, in fact, in the numerical examples with follow). Moreover, as we go on to demonstrate, this correction results in the preservation of both of these quantities, to first order, and offers some consequent improvement in the numerical behaviour of the energy minimisation scheme.

## 5.6 Tensorial Consistency in Energy Gradients

In this section we briefly investigate the non-linear conjugate gradients energy minimisation scheme used in the ONETEP method. In particular, we discuss what procedures are needed, if any, to ensure the tensorial validity of the energy gradients with respect to tangent and cotangent vectors of the support manifold and, in particular, compliance with geometrically induced corrections to the density kernel.

Energy minimisation is carried out in ONETEP with respect to the support functions within the assumption that the density kernel has no explicit dependence on the support functions or their duals. Geometrically, we would interpret this assumption as one whereby the density kernel does not change with respect to the local frame formed by the dual functions. In effect, the support functions and density kernel are thus treated as having no explicit inter-dependence, so that the total energy can be written as a separable functional of these two quantities. The energy is minimised with respect to the support functions in an outer conjugate gradients search loop with non-linear trial steps, using an appropriately scaled contravariant gradient

$$\Delta\phi_\alpha = \lambda g_\alpha = \lambda \left. E^{;\beta} \right|_{K^{\bullet\bullet}_{;\bullet} = 0 K^{\bullet\bullet}_{,\bullet} = 0} S_{\beta\alpha}. \tag{5.50}$$

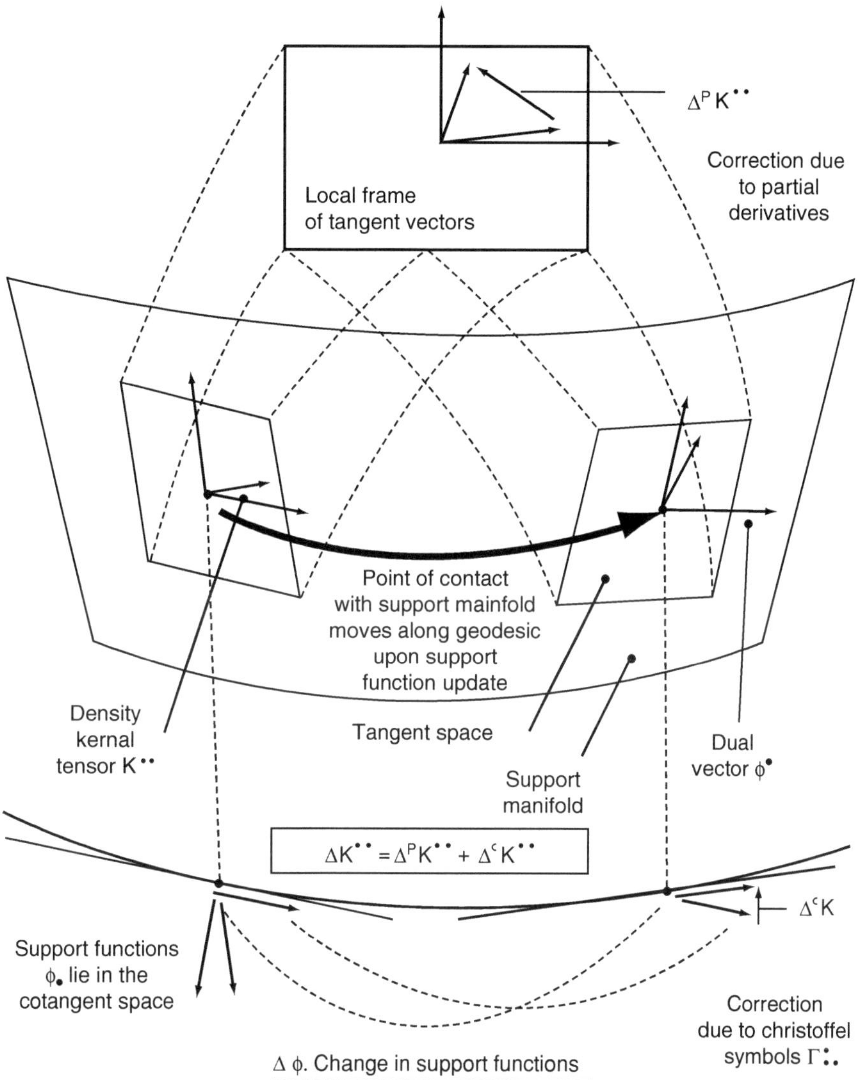


**Fig. 5.1** An interpretation of the components making up geometric corrections to the density kernel. Changes in the support functions cause a concomitant alteration to the point of contact of the tangent space with the support manifold. For a given density-matrix operator, this induces two separate correction terms to the density kernel. The first, due to partial derivatives of the density kernel, is associated with changes among the dual functions spanning the tangent space. The second correction, necessary due to the curvature of the support manifold, is introduced by the Christoffel symbols. This latter correction ensures that the density kernel retains its tensorial character in the new frame of support functions

This technique employs both the gradient and the evaluation of the energy at a number of trial steps, together with the history of the support function optimisation (the support function update vector must be conjugate to the space spanned by the previous update vectors), in order to predict the optimal step length $\lambda$.

Since no explicit dependence between kernel and support functions is included in the energy gradients, it is therefore inappropriate to include the $\Delta^{\mathcal{P}} K^{\bullet\bullet}$ and $\Delta^{\mathcal{P}} H_{\bullet\bullet}$ terms in the corrections to the density kernel and Hamiltonian, respectively. This latter observation is borne out in our numerical study detailed in Chap. 6. On the contrary, however, one cannot constrain the frame formed by the support functions and their duals in computing the energy gradient, by its very definition, and thus the inclusion of the Christoffel symbol terms $\Delta^{\mathcal{C}} K^{\bullet\bullet}$ and $\Delta^{\mathcal{C}} H_{\bullet\bullet}$ may not be obviated.

Since, by constraining the metric to preserve the completeness of the support manifold, we attach an explicit dependence of the density kernel on the support functions, it is natural to suggest an alternative approach whereby the assumption of independence between density kernel and functions is dropped, the explicit change in the kernel with the functions is included in the function gradient, the matrix updates $\Delta^{\mathcal{P}}$ being incorporated accordingly.

Such a scheme appears to be perhaps better controlled and more in keeping with more conventional cubic-scaling approaches to Kohn–Sham DFT, since in this way it is the density matrix operator itself (i.e., the single-particle states and their occupancies), and not the density kernel, which remains unaltered to first order during support function optimisation. We have not explored this possibility, however, as it would necessitate rather involved alterations to the methodology.

Let us assume, for simplicity, that the conventional ONETEP approach is used and that no geometric correction is applied to the Hamiltonian matrix, i.e., that it is fully reconstructed upon function update, but that a compensatory geometric update $\Delta K^{\bullet\bullet}$ to the density may be implied by the support function update. Although we explicitly constrain the partial derivative of the density kernel with respect to support functions to zero, there remains the contribution from the Christoffel symbols, $\Delta^{\mathcal{C}} K^{\bullet\bullet}$, and it is necessary to investigate whether this term adds an extra contribution to the support function energy gradient.

The scalar invariance of the energy means that no Christoffel symbol terms enter into its partial gradient or tangent vectors with respect to the support functions, thus the partial derivative of the energy is equivalent to the absolute derivative, and that derivative commutes with the metric tensor. Nonetheless, changes in the density kernel implied by support function update should be taken into account in the total derivatives of the energy. The necessary expression for the energy derivative is therefore generally given by

$$\lambda g_{\alpha} = \lambda \left( E^{;\beta} \Big|_{\substack{K^{\bullet\bullet;\bullet} = 0 \\ K^{\bullet\bullet,\bullet} = 0}} + \frac{\partial E}{\partial K^{\gamma\delta}} \left[ K^{\gamma\delta;\beta} \right]_{\substack{K^{\bullet\bullet;\bullet} = 0 \\ K^{\bullet\bullet,\bullet} = 0}} \right) S_{\beta\alpha}$$

$$
\begin{aligned}
&= \lambda \left( E^{;\beta}\Big|_{\substack{K^{\bullet\bullet}_{;\bullet}=0\\ K^{\bullet\bullet,\bullet}=0}} S_{\beta\alpha} + \frac{\partial E}{\partial K^{\gamma\delta}} \left[ K^{\gamma\delta}_{;\alpha} \right]_{\substack{K^{\bullet\bullet}_{;\bullet}=0\\ K^{\bullet\bullet,\bullet}=0}} \right) \\
&= y\lambda \left( E^{,\beta}\Big|_{\substack{K^{\bullet\bullet}_{;\bullet}=0\\ K^{\bullet\bullet,\bullet}=0}} S_{\beta\alpha} + \frac{\partial E}{\partial K^{\gamma\delta}} \begin{bmatrix} K^{\gamma\delta}_{,\alpha} \\ +K^{\epsilon\delta}\Gamma^{\gamma}_{\epsilon\alpha} \\ +K^{\gamma\epsilon}\Gamma^{\delta}_{\epsilon\alpha} \end{bmatrix}_{\substack{K^{\bullet\bullet}_{;\bullet}=0\\ K^{\bullet\bullet,\bullet}=0}} \right).
\end{aligned} \tag{5.51}
$$

The first term in the above is the usual corrected derivative of the energy with respect to the support functions, while those on the right are due to the concomitant change to the density kernel which automatically occurs upon support function update. We find that the two lattermost terms, involving Christoffel symbols, together vanish when subject to the constraint that there is no explicit dependence of the density kernel on the support functions, i.e.,

$$
\begin{aligned}
&\left[ K^{\epsilon\delta}\Gamma^{\gamma}_{\epsilon\alpha} + K^{\gamma\epsilon}\Gamma^{\delta}_{\epsilon\alpha} \right]_{K^{\bullet\bullet,\bullet}=0} \\
&= \left[ -\phi^{\gamma} S_{\epsilon\alpha} K^{\epsilon\delta} - \phi^{\delta} S_{\epsilon\alpha} K^{\gamma\epsilon} \right]_{K^{\bullet\bullet,\bullet}=0} \\
&= -\left[ \phi^{\gamma} K^{\delta\epsilon} S_{\epsilon\alpha} + \phi^{\delta} K^{\gamma\epsilon} S_{\epsilon\alpha} \right]_{K^{\bullet\bullet,\bullet}=0} \\
&= -\left[ K^{\gamma\delta,\epsilon} S_{\epsilon\alpha} \right]_{K^{\bullet\bullet,\bullet}=0} = 0.
\end{aligned} \tag{5.52}
$$

In addition, the contribution due to the partial derivative of the density kernel with respect to the dual functions, remembering that Ricci's Lemma does not apply to this term, provides a vanishing correction when subject to the constraint that there is no explicit dependence of the density kernel on the support functions, vis the expression

$$
\left[ K^{\gamma\delta}_{,\alpha} \right]_{K^{\bullet\bullet}_{\bullet\bullet,\bullet}=0} = \left[ \phi_{\epsilon} \left( K^{\epsilon\delta}\delta^{\gamma}_{\alpha} + K^{\epsilon\gamma}\delta^{\delta}_{\alpha} \right) \right]_{K^{\bullet\bullet}_{\bullet\bullet,\bullet}=0} = 0. \tag{5.53}
$$

Returning, finally, to our expression for the energy gradient, we find that it is not augmented by any terms in order to account for the geometrically-induced change in the density kernel with respect to the support functions, and simply remains

$$
\lambda g_{\alpha} = \lambda\ E^{,\beta}\Big|_{K^{\bullet\bullet}_{\bullet\bullet,\bullet}=0 K^{\bullet\bullet,\bullet}=0} S_{\beta\alpha}. \tag{5.54}
$$

We conclude that, at least for the manifold compatible with preservation of the completeness of the space spanned by the support functions and ignoring changes induced in the Hamiltonian matrix, no correction is needed in the derivative of the energy with respect to the support functions in order to make the derivative consistent with the change in energy induced by compensatory updates to the density kernel.

Let us next make some observations with regard to geometric issues surrounding the non-linear step which is incorporated in the conjugate gradients energy minimisation algorithm. We consider the simplest, namely quadratic, step length extrapolation scheme. In this method, the energy $E^{(0)}$ and gradient $g^{(0)}_{\bullet}$ for a given set of support functions are first computed with an optimised density kernel, after which a step is taken in the covariant gradient direction using a step length $\lambda^{(1)}$ computed using the history of the energy minimisation moves. The energy is then recomputed at this trial step, giving a value $E^{(1)}$. A concave parabola is fitted through the energies $E^{(0)}$ and $E^{(1)}$, given the slope $g_{\bullet}^{(0)}$ and the step length $\lambda^{(1)}$, and the selected step length $\lambda^{(2)}$ is then that which brings us to the predicted minimum $E^{(2)}$ of this parabola.

A noteworthy point on the coupling of the non-linear conjugate gradients algorithm to geometrically induced changes to tensors such as the density kernel or Hamiltonian is that it is generally insufficient to apply the geometric corrections only after the algorithm has selected an optimal support function step length $\lambda^{(2)}$. In order to ensure the consistency between gradients and energy differences in the conjugate gradients step-length selection scheme, it is necessary to apply appropriate trial corrections to the density kernel and Hamiltonian which are commensurate with each trial update vector, otherwise the step-length prediction method is not able to take account of the geometric transport of these tensors. This point is illustrated for the case of density kernel update in a numerical study described in Chap. 6, where it is shown that little or no improvement in total energy convergence is offered by the geometric corrections unless they are fully integrated with the non-linear conjugate gradients algorithm.

There is a further subtlety to such methods, however, which we mention but do not explore in detail. Let us re-examine our simple example of a quadratic step prediction scheme, and suppose that we have carried it out for a given set of support functions along with their accompanying density kernel. If we return to the original point $E^{(0)}$, but instead take a trial step with length $\lambda^{(1*)} = \lambda^{(2)} - \lambda^{(1)}$, we may compute an alternative trial energy $E^{(1*)}$ and, as before, calculate an alternative predicted step length $\lambda^{(2*)} \approx \lambda^{(1)}$ and energy $E^{(2*)}$ using a slightly different parabola. The difficulty here is that, quite aside from the possible non-parabolicity of the energy landscape, there will generally be a discrepancy the between the energies $E^{(2)}$ and $E^{(2*)}$, and consequently also between the predicted step lengths $\lambda^{(2)}$ and $\lambda^{(2*)}$. This discrepancy is due the non-triviality of the Riemann-Christoffel tensor, which expresses the non-commutativity of parallel transport via different paths on the curved support manifold. Thus, the optimisation algorithm may not be optimal, in general, for the curved nature of the manifold. These topics are touched upon further in Appendix A.3, though a full investigation into how to provide a robust improvement upon the algorithm, perhaps using a geodesic measure of the step length, remains an avenue for future work.

## 5.7 First-Order Density-Matrix Preservation

In this section, we investigate the preservation of certain important properties of the density kernel under support function update when the kernel geometric corrections are invoked. We will focus on the density kernel correction terms of importance in the ONETEP method which, due to the treatment of the density kernel as explicitly independent of the support functions, are those proportional to Christoffel symbols only. This investigation provides a motivation for the analysis of the changes observed in the numerical behaviour of the method, when corrections are applied, described in Chap. 6.

In order to proceed, let us consider three important properties which must be satisfied by the density kernel and the Hamiltonian matrix corresponding to the ground state. Firstly, the density-matrix must be idempotent if it is to describe fermionic orthogonal single-particle orbitals. Usually, the idempotency of the density-matrix is not preserved under changes in support functions in any sense. Assuming that the idempotency deviation is null before the support function update, the deviation thereby introduced is given by

$$\begin{aligned} &|\phi_\alpha + \Delta\phi_\alpha\rangle K^{\alpha\beta} \langle\phi_\beta + \Delta\phi_\beta|\phi^\gamma + \Delta\phi^\gamma\rangle \\ &\quad \times S_{\gamma\delta} K^{\delta\epsilon} S_{\epsilon\zeta} \langle\phi^\zeta + \Delta\phi^\zeta| - |\phi_\alpha\rangle K^{\alpha\beta} \langle\phi_\beta|\phi^\gamma\rangle S_{\gamma\delta} K^{\delta\epsilon} S_{\epsilon\zeta} \langle\phi^\zeta| \end{aligned} \tag{5.55}$$

and, typically, a number of iterations of the third-order purifying transformation

$$K'^{\bullet\bullet} = 3K^{\bullet\bullet} S_{\bullet\bullet} K^{\bullet\bullet} - 2K^{\bullet\bullet} S_{\bullet\bullet} K^{\bullet\bullet} S_{\bullet\bullet} K^{\bullet\bullet} \tag{5.56}$$

are required to bring the density matrix back to a sufficiently idempotent state for the kernel optimisation to proceed. The purifying transformations require a non-negligible cost to compute but, more importantly, introduce some spoilage to the density; consequently implying additional effort in inner-loop optimisation in order to restore the density-matrix to its previous quality. The second property that is, in general, broken by support function update is the compatibility of the density-matrix and the Hamiltonian, that is the multiplicative commutativity of these two operators. Assuming that the compatibility holds before the support functions change (for the sake of illustration), the error introduced is given generally by

$$\begin{aligned} &|\phi^\alpha + \Delta\phi^\alpha\rangle H_{\alpha\beta} K^{\beta\gamma} \langle\phi_\gamma + \Delta\phi_\gamma| \\ &\quad - |\phi_\alpha + \Delta\phi_\alpha\rangle K^{\alpha\beta} H_{\beta\gamma} \langle\phi^\gamma + \Delta\phi^\gamma| \end{aligned} \tag{5.57}$$

and, in practice, it makes up an appreciable fraction of the density matrix corruption which must be first recovered by the LNVD method [19–21] before any improvement in the density can be obtained from the improved support functions. Finally, of course, the density kernel must describe the correct number of electrons in the system. No attempt is made to make sure that the support functions are in any sense normalised—it would be redundant to do so for nonorthogonal functions—and the occupancy is absorbed into the density kernel. Following support function update, assuming that the density kernel is first well normalised, the resulting deviation from the correct electron number is described by

$$K^{\alpha\beta} \langle\phi_\beta + \Delta\phi_\beta|\phi_\alpha + \Delta\phi_\alpha\rangle - K^{\alpha\beta} \langle\phi_\beta|\phi_\alpha\rangle. \tag{5.58}$$

Let us now re-evaluate the same expressions for the first order error in the idempotency, compatibility and normalisation, applying the corrections to the density kernel and Hamiltonian provided by the terms in the absolute derivative proportional to Christoffel symbols. To first order in the update vector, and noting that the density kernel correction itself is a first order term in the update vector, the revised error in the idempotency is given by the expression

$$\begin{aligned} &|\phi_\alpha + \Delta\phi_\alpha\rangle \left(K + \Delta^C K\right)^{\alpha\beta} \langle\phi_\beta + \Delta\phi_\beta|\phi^\gamma + \Delta\phi^\gamma\rangle S_{\gamma\delta} \\ &\times \left(K + \Delta^C K\right)^{\delta\epsilon} S_{\epsilon\zeta} \langle\phi^\zeta + \Delta\phi^\zeta| - |\phi_\alpha\rangle K^{\alpha\beta} \langle\phi_\beta|\phi^\gamma\rangle S_{\gamma\delta} K^{\delta\epsilon} S_{\epsilon\zeta} \langle\phi^\zeta| \end{aligned} \tag{5.59}$$

which approximates, neglecting all but first-order terms in the support function change vector, to

$$\begin{aligned} &|\phi_\alpha\rangle \left(\Delta^C K\right)^{\alpha\beta} \langle\phi_\beta|\phi^\gamma\rangle S_{\gamma\delta} K^{\delta\epsilon} S_{\epsilon\zeta} \langle\phi^\zeta| \\ &+ |\phi_\alpha\rangle K^{\alpha\beta} \langle\phi_\beta|\phi^\gamma\rangle S_{\gamma\delta} \left(\Delta^C K\right)^{\delta\epsilon} S_{\epsilon\zeta} \langle\phi^\zeta| \\ &+ |\Delta\phi_\alpha\rangle K^{\alpha\beta} \langle\phi_\beta|\phi^\gamma\rangle S_{\gamma\delta} K^{\delta\epsilon} S_{\epsilon\zeta} \langle\phi^\zeta| \\ &+ |\phi_\alpha\rangle K^{\alpha\beta} \langle\Delta\phi_\beta|\phi^\gamma\rangle S_{\gamma\delta} K^{\delta\epsilon} S_{\epsilon\zeta} \langle\phi^\zeta| \\ &+ |\phi_\alpha\rangle K^{\alpha\beta} \langle\phi_\beta|\Delta\phi^\gamma\rangle S_{\gamma\delta} K^{\delta\epsilon} S_{\epsilon\zeta} \langle\phi^\zeta| \\ &+ |\phi_\alpha\rangle K^{\alpha\beta} \langle\phi_\beta|\phi^\gamma\rangle S_{\gamma\delta} K^{\delta\epsilon} S_{\epsilon\zeta} \langle\Delta\phi^\zeta|. \end{aligned} \tag{5.60}$$

This operator in fact vanishes, since it reduces to

$$\begin{aligned} &|\phi_\alpha\rangle \left(-S^{\alpha\zeta} \langle\phi_\zeta|\Delta\phi_\gamma\rangle K^{\gamma\beta} - K^{\alpha\gamma} \langle\Delta\phi_\gamma|\phi_\zeta\rangle S^{\zeta\beta}\right) S_{\beta\delta} K^{\delta\epsilon} \langle\phi_\epsilon| \\ &+ |\phi_\alpha\rangle K^{\alpha\beta} S_{\beta\delta} \left(-S^{\delta\zeta} \langle\phi_\zeta|\Delta\phi_\gamma\rangle K^{\gamma\epsilon} - K^{\delta\gamma} \langle\Delta\phi_\gamma|\phi_\zeta\rangle S^{\zeta\epsilon}\right) \langle\phi_\epsilon| \\ &+ |\Delta\phi_\alpha\rangle K^{\alpha\beta} S_{\beta\delta} K^{\delta\epsilon} \langle\phi_\epsilon| \\ &+ |\phi_\alpha\rangle K^{\alpha\beta} \langle\Delta\phi_\beta|\phi_\delta\rangle K^{\delta\epsilon} \langle\phi_\epsilon| \\ &+ |\phi_\alpha\rangle K^{\alpha\beta} \langle\phi_\beta|\Delta\phi_\delta\rangle K^{\delta\epsilon} \langle\phi_\epsilon| \\ &+ |\phi_\alpha\rangle K^{\alpha\beta} S_{\beta\delta} K^{\delta\epsilon} \langle\Delta\phi_\epsilon| = 0. \end{aligned} \tag{5.61}$$

We turn next to the first order error introduced to the compatibility condition. By necessity, since the kinetic energy, for example, is explicitly dependent on the form of the support functions, the Hamiltonian operator is fully recomputed whenever the support functions are updated. First order changes to the Hamiltonian operator could, in principle, actually be calculated by explicitly considering explicit definitions of the Hamiltonian, though this is rather beyond the scope of this study. We neglect perturbations to the Hamiltonian operator here and consider only the geometrically-induced changes to the compatibility.

It is of interest, however, to note that first order change in the commutator between the density kernel and the Hamiltonian matrix computed using the Christoffel symbol induced corrections, given by

$$
\begin{aligned}
&|\phi^{\alpha} + \Delta\phi^{\alpha}\rangle \left(H + \Delta^{C} H\right)_{\alpha\beta} \left(K + \Delta^{C} K\right)^{\beta\gamma} \langle\phi_{\gamma} + \Delta\phi_{\gamma}| \\
&\quad - |\phi_{\alpha} + \Delta\phi_{\alpha}\rangle \left(K + \Delta^{C} K\right)^{\alpha\beta} \left(H + \Delta^{C} H\right)_{\beta\gamma} \langle\phi^{\gamma} + \Delta\phi^{\gamma}| \\
&\quad - |\phi^{\alpha} + \Delta\phi^{\alpha}\rangle H_{\alpha\beta} K^{\beta\gamma} \langle\phi_{\gamma} + \Delta\phi_{\gamma}| \\
&\quad\; + |\phi^{\alpha} + \Delta\phi^{\alpha}\rangle K^{\alpha\beta} H_{\beta\gamma} \langle\phi^{\gamma} + \Delta\phi^{\gamma}|,
\end{aligned} \tag{5.62}
$$

which expands to, neglecting second and higher-order terms in the support function update vector,

$$
\begin{aligned}
&|\phi^{\alpha}\rangle \left(\langle\Delta\phi_{\alpha}|\phi_{\zeta}\rangle S^{\zeta\eta} H_{\eta\beta} + H_{\alpha\eta} S^{\eta\zeta} \langle\phi_{\zeta}|\Delta\phi_{\beta}\rangle\right) K^{\beta\gamma} \langle\phi_{\gamma}| \\
&\quad - |\phi_{\alpha}\rangle K^{\alpha\beta} \left(\langle\Delta\phi_{\beta}|\phi_{\zeta}\rangle S^{\zeta\eta} H_{\eta\gamma} + H_{\beta\eta} S^{\eta\zeta} \langle\phi_{\zeta}|\Delta\phi_{\gamma}\rangle\right) \langle\phi^{\gamma}| \\
&\quad + |\phi^{\alpha}\rangle H_{\alpha\beta} \left(-S^{\beta\zeta} \langle\phi_{\zeta}|\Delta\phi_{\eta}\rangle K^{\eta\gamma} - K^{\beta\eta} \langle\Delta\phi_{\eta}|\phi_{\zeta}\rangle S^{\zeta\gamma}\right) \langle\phi_{\gamma}| \\
&\quad - |\phi_{\alpha}\rangle \left(-S^{\alpha\zeta} \langle\phi_{\zeta}|\Delta\phi_{\eta}\rangle K^{\eta\beta} - K^{\alpha\eta} \langle\Delta\phi_{\eta}|\phi_{\zeta}\rangle S^{\zeta\beta}\right) H_{\beta\gamma} \langle\phi^{\gamma}|.
\end{aligned} \tag{5.63}
$$

This expression may, in turn, be broken up into the fragments

$$
\begin{aligned}
&|\phi^{\alpha}\rangle\langle\Delta\phi_{\alpha}|\phi_{\zeta}\rangle S^{\zeta\eta} H_{\eta\beta} K^{\beta\gamma} \langle\phi_{\gamma}| + |\phi^{\alpha}\rangle H_{\alpha\eta} S^{\eta\zeta} \langle\phi_{\zeta}|\Delta\phi_{\beta}\rangle K^{\beta\gamma} \langle\phi_{\gamma}| \\
&\quad - |\phi_{\alpha}\rangle K^{\alpha\beta} \langle\Delta\phi_{\beta}|\phi_{\zeta}\rangle S^{\zeta\eta} H_{\eta\gamma} \langle\phi^{\gamma}| - |\phi_{\alpha}\rangle K^{\alpha\beta} H_{\beta\eta} S^{\eta\zeta} \langle\phi_{\zeta}|\Delta\phi_{\gamma}\rangle \langle\phi^{\gamma}| \\
&\quad - |\phi^{\alpha}\rangle H_{\alpha\beta} \langle\phi^{\beta}|\Delta\phi_{\eta}\rangle K^{\eta\gamma} \langle\phi_{\gamma}| - |\phi^{\alpha}\rangle H_{\alpha\beta} K^{\beta\eta} \langle\Delta\phi_{\eta}| \\
&\quad + |\Delta\phi_{\eta}\rangle K^{\eta\beta} H_{\beta\gamma} \langle\phi^{\gamma}| + |\phi_{\alpha}\rangle K^{\alpha\eta} \langle\Delta\phi_{\eta}|\phi^{\beta}\rangle H_{\beta\gamma} \langle\phi^{\gamma}|.
\end{aligned} \tag{5.64}
$$

Here we immediately see that some pairwise cancellation occurs, and the symmetry of the operator $|\phi^{\zeta}\rangle\langle\Delta\phi_{\zeta}| = |\Delta\phi_{\zeta}\rangle\langle\phi^{\zeta}| = |\Delta\phi^{\zeta}\rangle\langle\phi_{\zeta}|$ can be employed to further simplify the compatibility error to

$$
\begin{aligned}
&|\phi^{\alpha}\rangle\langle\Delta\phi_{\alpha}|\phi^{\eta}\rangle H_{\eta\beta} K^{\beta\gamma} \langle\phi_{\gamma}| - |\phi_{\alpha}\rangle K^{\alpha\beta} H_{\beta\eta} \langle\phi^{\eta}|\Delta\phi_{\gamma}\rangle \langle\phi^{\gamma}| \\
&\quad + |\Delta\phi_{\eta}\rangle K^{\eta\beta} H_{\beta\gamma} \langle\phi^{\gamma}| - |\phi^{\alpha}\rangle H_{\alpha\beta} K^{\beta\eta} \langle\Delta\phi_{\eta}| \\
&= |\Delta\phi^{\alpha}\rangle\langle\phi_{\alpha}|\phi^{\eta}\rangle H_{\eta\beta} K^{\beta\gamma} \langle\phi_{\gamma}| - |\phi_{\alpha}\rangle K^{\alpha\beta} H_{\beta\eta} \langle\phi^{\eta}|\phi_{\gamma}\rangle \langle\Delta\phi^{\gamma}| \\
&\quad + |\Delta\phi_{\eta}\rangle K^{\eta\beta} H_{\beta\gamma} \langle\phi^{\gamma}| - |\phi^{\alpha}\rangle H_{\alpha\beta} K^{\beta\eta} \langle\Delta\phi_{\eta}| \\
&= |\Delta\phi^{\eta}\rangle H_{\eta\beta} K^{\beta\alpha} \langle\phi_{\alpha}| - |\phi_{\alpha}\rangle K^{\alpha\beta} H_{\beta\eta} \langle\Delta\phi^{\eta}| \\
&\quad + |\Delta\phi_{\eta}\rangle K^{\eta\beta} H_{\beta\alpha} \langle\phi^{\alpha}| - |\phi^{\alpha}\rangle H_{\alpha\beta} K^{\beta\eta} \langle\Delta\phi_{\eta}|,
\end{aligned} \tag{5.65}
$$

which does not vanish, in general. Of course, since it is antisymmetric by construction, it possesses a null trace.

Finally, we note that the first order error introduced to the total occupancy of the system vanishes when the density kernel is appropriately corrected for the parallel transport of the support space, as we may deduce from

$$
\begin{aligned}
&\left(K + \Delta^C K\right)^{\alpha\beta} \langle \phi_\beta + \Delta\phi_\beta | \phi_\alpha + \Delta\phi_\alpha \rangle - K^{\alpha\beta} \langle \phi_\beta | \phi_\alpha \rangle \\
&= \left(-S^{\alpha\zeta} \langle \phi_\zeta | \Delta\phi_\eta \rangle K^{\eta\beta} - K^{\alpha\eta} \langle \Delta\phi_\eta | \phi_\zeta \rangle S^{\zeta\beta}\right) \langle \phi_\beta | \phi_\alpha \rangle + K^{\alpha\beta} \langle \Delta\phi_\beta | \phi_\alpha \rangle \\
&\quad + K^{\alpha\beta} \langle \phi_\beta | \Delta\phi_\alpha \rangle \\
&= - \langle \phi_\beta | \Delta\phi_\eta \rangle K^{\eta\beta} - K^{\alpha\eta} \langle \Delta\phi_\eta | \phi_\alpha \rangle + K^{\alpha\beta} \langle \Delta\phi_\beta | \phi_\alpha \rangle \\
&\quad + K^{\alpha\beta} \langle \phi_\beta | \Delta\phi_\alpha \rangle = 0.
\end{aligned} \tag{5.66}
$$

The preservation of the idempotency and trace of the density-matrix (the density-matrix is entirely preserved to first order, in fact) is a rather interesting property of the geometric correction. It arises, fundamentally, as a consequence of the assumption that the identity operator on the support space does not change to first-order with the support function update, and so the extent to which this preservation holds in practice must depend on the validity of this assumption.

These results provide a promising suggestion that the density kernel corrected using this technique may provide a superior starting point, after support function update, than the uncorrected kernel, leaving the energy minimisation algorithm in a better position to find the best density expressible in the updated support space. This hypothesis is explored in a detailed computational study in Chap. 6.

Our proposed geometric method for density matrix preservation under representation change may also prove to be useful in wider contexts, such as in the time-propagation of the representation for dynamical systems, such as in TDDFT, or, as we discuss in Chap. 7, in the ab initio study of strong correlation effects.

## 5.8 Concluding Remarks

To conclude, we have performed a detailed analysis of the geometric ramifications of using nonorthogonal sets of functions to express the single-particle states in ab initio methods. The Riemannian manifold to which those functions lie cotangent, the *support manifold* and its differential geometry, has not been previously studied to our knowledge. We have shown that when the support functions are altered according to any mechanism, for example energy minimisation, such geometric considerations naturally introduce corrective terms to tensors such as the density kernel and Hamiltonian.

Moreover, these corrective terms act to preserve the idempotency and trace of the density-matrix to first order. This provides a solution to the long-standing technical obstacle of idempotency corruption under representation update. The correction is trivially implemented in a modern linear-scaling total-energy approach and may prove to be rather valuable in accelerating and stabilising such methods.

# References


1. C.-K. Skylaris, A.A. Mostofi, P.D. Haynes, O. Diéguez, M.C. Payne, Nonorthogonal generalized Wannier function pseudopotential plane-wave method. Phys. Rev. B **66**(3), 035119 (2002)
2. M.J. Han, T. Ozaki, J. Yu, *O(N) LDA+U* electronic structure calculation method based on the nonorthogonal pseudoatomic orbital basis. Phys. Rev. B **73**(4), 045110 (2006)
3. E. Hernández, M.J. Gillan, Self-consistent first-principles technique with linear scaling. Phys. Rev. B **51**(15), 10157 (1995)
4. F. Mauri, G. Galli, Electronic-structure calculations and molecular-dynamics simulations with linear system-size scaling. Phys. Rev. B **50**(7), 4316 (1994)
5. S. Goedecker, Linear scaling electronic structure methods. Rev. Mod. Phys. **71**(4), 1085 (1999)
6. P. Hohenberg, W. Kohn, Inhomogeneous electron gas. Phys. Rev. **136**(3B), B864 (1964)
7. W. Kohn, L.J. Sham, Self-consistent equations including exchange and correlation effects. Phys. Rev. **140**(4A), A1133 (1965)
8. E. Runge, E.K.U. Gross, Density-functional theory for time-dependent systems. Phys. Rev. Lett. **52**(12), 997 (1984)
9. C.-K. Skylaris, P.D. Haynes, A.A. Mostofi, M.C. Payne, Introducing ONETEP: linear-scaling density functional simulations on parallel computers. J. Chem. Phys. **122**, 084119 (2005)
10. P.D. Haynes, C.-K. Skylaris, A.A. Mostofi, M.C. Payne, Elimination of basis set superposition error in linear-scaling density-functional calculations with local orbitals optimised in situ. Chem. Phys. Lett. **422**, 345 (2006)
11. F. Mauri, G. Galli, R. Car, Orbital formulation for electronic-structure calculations with linear system-size scaling. Phys. Rev. B **47**(15), 9973 (1993)
12. J. Kim, F. Mauri, G. Galli, Total-energy global optimizations using nonorthogonal localized orbitals. Phys. Rev. B **52**(3), 1640 (1995)
13. P. Ordejón, D.A. Drabold, R.M. Martin, M.P. Grumbach, Linear system-size scaling methods for electronic-structure calculations. Phys. Rev. B **51**(3), 1456 (1995)
14. B. Spain, *Tensor Calculus - A Concise Course*. (Dover, New York, 2003)
15. M. Deserno, Notes on differential geometry (2004), http://www.cmu.edu/biolphys/deserno/pdf/diff_geom.pdf.
16. D. Lovelock, H. Rund, *Tensors, Differential Forms and Variational Principles*. (Dover, New York, 1989)
17. A. Das, *Tensors—The Mathematics of Relativity Theory and Continuum Mechanics*. (Springer, New York, 2007)
18. D.J. Hurley, M.A. Vandyck, *Geometry, Spinors and Applications*. (Springer-Praxis, Berlin, 2000)
19. X.-P. Li, R.W. Nunes, D. Vanderbilt, Density-matrix electronic-structure method with linear system-size scaling. Phys. Rev. B **47**(16), 10891 (1993)
20. R.W. Nunes, D. Vanderbilt, Generalization of the density-matrix method to a nonorthogonal basis. Phys. Rev. B **50**(23), 17611 (1994)
21. M.S. Daw, Model for energetics of solids based on the density matrix. Phys. Rev. B **47**(16), 10895 (1993)

# Chapter 6
# A Numerical Study of Geometric Corrections for Representation Optimisation

The parallel transport of tensors such as the density kernel and Hamiltonian has been shown in the previous chapter to contribute non-zero correction terms when the *support functions* representing single-particle orbitals in density-matrix based ab initio calculations are allowed to change.

Since support function optimisation is a characteristic feature of linear-scaling methods which seek to reproduce the accuracy of their cubic-scaling plane-wave counterparts [1–6], it is particularly important for us to investigate whether such geometric corrections are a help or a hindrance in realistic total-energy calculations. We rather take the view that, since these corrections were shown in the previous chapter to be unambiguously required by the formalism, if their use were to prove counterproductive in practice then we would suspect the superior performance observed in their neglect to be due to a somewhat unreliable cancellation of errors. Fortuitously, however, we have found that, for the systems studied and using the corrections appropriate to the method in use, convergence may be substantially improved in both in speed and stability. At worst, convergence is neither significantly ameliorated or impaired.

## 6.1 Computational Methodology for Naphthalene

With a view to elucidating the effect of the geometrically-derived corrections to the density kernel as it evolves in a changing frame of support functions, we performed a numerical study of the convergence to the ground state of the naphthalene molecule, a small polycyclic aromatic hydrocarbon with chemical formula $C_{10}H_8$ using the ONETEP method [1, 2]. The support functions constructed during total energy minimisation in this method are known as nonorthogonal generalised Wannier functions (NGWFs) [7].

The gas-phase naphthalene molecule and the longer oligoacene molecules studied later in this chapter were chosen because they have been previously particularly well studied with first principles methods (particularly naphthalene), see for example

D. D. O'Regan, *Optimised Projections for the Ab Initio Simulation of Large and Strongly Correlated Systems*, Springer Theses, DOI: 10.1007/978-3-642-23238-1_6,

Refs. [8–10] and references therein, and because the conventional energy minimisation scheme in this code already performs rather well for these systems.

The Hamiltonian matrix was fully re-computed upon alteration of either the NGWFs or the density kernel, as is ordinarily the case in ONETEP, as an investigation of the perturbative treatment of the Hamiltonian, while worthy, would complicate the analysis of the parallel transport corrections to the density kernel. The separate terms making up the absolute differential of the density kernel, due to partial derivatives and Christoffel symbols, were treated separately in order to confirm our understanding of which contributions are those appropriate to the ONETEP approach. The effect of including, or neglecting, the corrective terms to the density kernel within the NGWF non-linear step-length optimisation scheme (the Fletcher-Reeves algorithm) was also separately analysed.

A *psinc* function [11, 12] grid spacing equivalent to a plane-wave energy cutoff of between 1025 eV and 1135 eV (the *basis*, in our nomenclature, has a finesse which, for technical reasons, is anisotropic when a non-cubic simulation cell is used) was used with NGWFs, four on each carbon atom, one on each hydrogen, truncated to a sphere of radius 10 $a_0$ (the *support functions*) to represent the Kohn–Sham density-matrix. No truncation of the density kernel was applied in the case of naphthalene, the NGWF overlap matrix was inverted exactly where necessary and the molecular geometry was optimised (with no geometric corrections) in a fixed supercell large enough to ensure no overlap between NGWFs and their periodic images ($32\,a_0 \times 28\,a_0 \times 22\,a_0$).

The PBE generalised gradient [13] exchange correlation functional was used throughout and included in the construction of the pseudopotentials using the OPIUM code.[1] A convergence threshold of $2 \times 10^{-7}$ Ha $a_0^{3/2}$ was used for the root mean squared NGWF gradient, with a maximum step length of $4.0\,a_0^{-3/2}$, and the complete history of the conjugate gradients steps was retained.

The density kernel was optimised for each set of NGWFs using a combination of the LNVD [14–16] and penalty functional [17] methods, between 10 and 15 iterations of the former were used on each step. A maximum kernel step length of 6.0 e was allowed, a convergence threshold of $1 \times 10^{-10}$ Ha $e^{-1}$ was used, and no re-optimisation of the density kernel (by LNVD or penalty functional methods) was performed at NGWF trial steps.

Geometric corrections were applied only to the density kernel, as stated, and the Hamiltonian matrix was re-evaluated completely upon each change in NGWFs or density kernel.

[1] A set of RRKJ Pseudopotentials were generated using the Opium code, http://opium.sourceforge.net, using the GGA input parameters available therein, optimized for a minimum plane-wave cutoff of 680 eV, albeit with a scalar-relativistic correction for all species and, for the transition-metal ions, some slight modifications to the core radii and a non-linear core correction of Fuchs-Scheffler characteristic radius 1.3a.u.

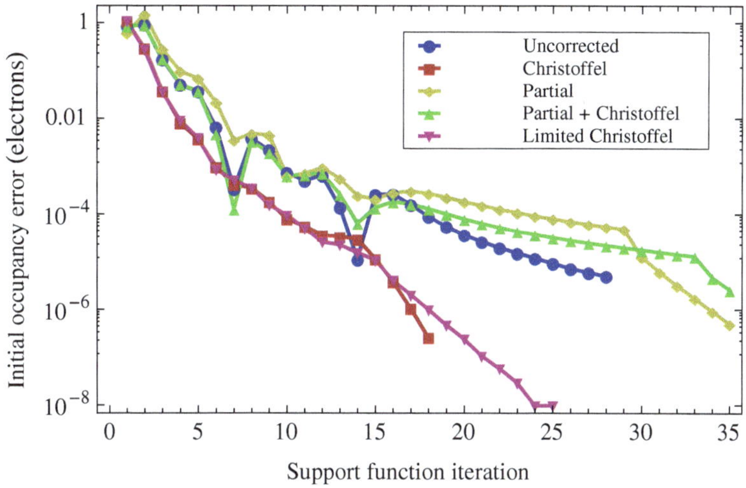


**Fig. 6.1** The absolute value of the deviation of the total occupancy from its ideal value, computed immediately after NGWF update, as a function of the NGWF update iteration. Behaviour is shown for the case where no correction is applied to the density kernel upon function change, and when combinations of corrections corresponding to the partial and geometric terms of the absolute derivative are performed

## 6.2 Geometric Density Kernel Corrections in Naphthalene

Figure 6.1 shows the deviation of the total occupancy of the system from its ideal value (48 e) immediately after NGWF update, as a function of the NGWF update iteration number. We observe that in this regard, the corrections given by the metric-corrected covariant partial derivative ($\Delta^{\mathcal{P}} K^{\bullet\bullet}$—denoted PARTIAL and provided in Eq. 5.46 ) and that combined with the correction induced by the Christoffel symbol ($\Delta^{\mathcal{P}} K^{\bullet\bullet} + \Delta^{\mathcal{C}} K^{\bullet\bullet}$—denoted PARTIAL + CHRISTOFFEL, given by Eq. 5.49 ) perform worse as compared to the uncorrected density kernel (denoted UNCORRECTED). On the other hand, when the Christoffel-induced correction terms alone are used ($\Delta^{\mathcal{C}} K^{\bullet\bullet}$—denoted CHRISTOFFEL, see Eq. 5.48 ) the occupancy error suffered is at almost all stages substantially lower than in the uncorrected method, exhibiting rather less erratic behaviour.

The CHRISTOFFEL correction, and that correction alone, is required when the stationarity of the density kernel with respect to the local cotangent space to the support manifold is assumed during computation of the energy gradient with respect to the NGWFs, as described in Sect. 5.6 and as is always the case in ONETEP at present. The terms proportional to Christoffel symbols correct for the rocking of the tangent space on the support manifold. The observed improvement in the behaviour of the occupancy when the CHRISTOFFEL correction is used accords well with our previous analytical result that the occupancy error after NGWF update should vanish to first order in the NGWF step length with this method.

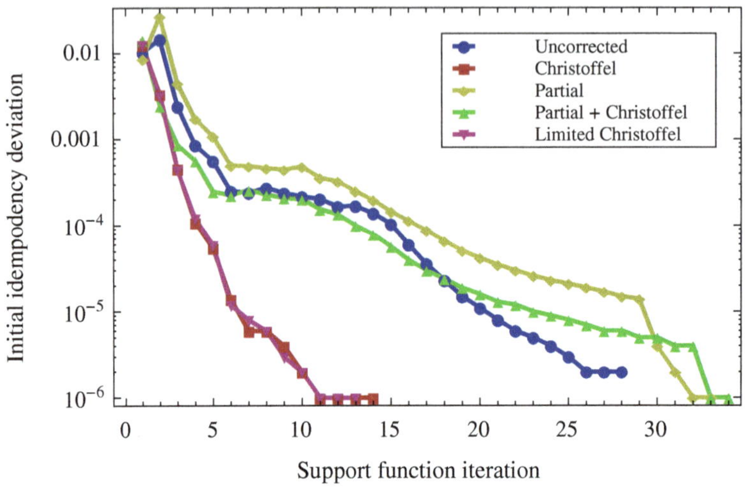


**Fig. 6.2** The value of the density kernel idempotency error, computed immediately after NGWF update, as a function of the NGWF update iteration. Behaviour is shown for the case where no correction is applied to the density kernel upon function change, and when combinations of corrections corresponding to the partial and geometric terms of the absolute derivative are performed

This condition holds true also for the restriction of the CHRISTOFFEL method to application only after the optimal NGWF update step has been selected, and not at trial NGWF steps in the non-linear conjugate gradients algorithm, an incomplete correction which we denote LIMITED CHRISTOFFEL. The observed similarity between the latter two methods, CHRISTOFFEL and LIMITED CHRISTOFFEL, in terms of occupancy preservation is unsurprising since the occupancy error is evaluated immediately after the correction which is common to both.

The trace of the idempotency deviation operator, $Tr\left[\hat{\rho}\hat{\rho} - \hat{\rho}\right]$, computed immediately after NGWF update, is depicted in Fig. 6.2. In this case, we see that the beneficial effect of the CHRISTOFFEL geometric correction, $\Delta^{\mathcal{C}} K^{\bullet\bullet}$, is more pronounced than for the absolute occupancy, indeed towards the end of the calculation the initial idempotency error is reduced by approximately two orders of magnitude using this correction (so much so that one could feasibly obviate penalty-functional type idempotency corrections entirely towards the latter stages, with a concomitant saving in computational overhead).

Conversely, the inclusion of the PARTIAL correction, $\Delta^{\mathcal{P}} K^{\bullet\bullet}$, which is not expected to be appropriate for this energy minimisation algorithm, tends to somewhat disimprove the idempotency of the density matrix and necessitate more NGWF update steps before convergence is achieved.

As a final point on this topic, we note that the formula for the CHRISTOFFEL correction, Eq. 5.48 , is somewhat reminiscent of a well-known result in density-matrix theory, first demonstrated in Ref. [17], which states that changes to the density-matrix $\hat{\rho}$ which preserve its idempotency to first order must be expressible in the form

$$\left(\hat{1} - \hat{\rho}\right) \hat{\Delta}\hat{\rho} + \hat{\rho}\hat{\Delta}^{\dagger} \left(\hat{1} - \hat{\rho}\right). \tag{6.1}$$

As a result of this similarity, and for the sake of thoroughness, we investigated numerous density kernel updates of this form. We found that the correction term

$$\begin{aligned} \Delta^{\mathcal{M}} K^{\alpha\beta} = & - (S - K)^{\alpha\gamma} \langle\phi_{\gamma}|\Delta\phi_{\delta}\rangle K^{\delta\beta} \\ & - K^{\alpha\delta} \langle\Delta\phi_{\delta}|\phi_{\gamma}\rangle (S - K)^{\gamma\beta} \end{aligned} \tag{6.2}$$

provided the best performance of those investigated in our tests on naphthalene but that it did not, being derived for stationary NGWFs, offer the improvements yielded by $\Delta^{\mathcal{C}} K^{\bullet\bullet}$ in terms of the density-matrix preservation under NGWF update seen in Figs. 6.1 and 6.2.

It may be tempting, in this context, to view the density kernel corrections in terms of the theory of excitations in the density matrix. Formally, given a fixed set of support functions, we think of the density kernel change $\Delta^{\mathcal{M}} K^{\bullet\bullet}$ as one describing single-particle excitations, of probability $\langle\Delta\phi_{\delta}|\phi_{\gamma}\rangle$, between the space of occupied orbitals described by $K^{\bullet\bullet}$ and the space of virtual orbitals described by $(S - K)^{\bullet\bullet}$. Accordingly, the terms $\Delta^{\mathcal{P}} K^{\bullet\bullet}$ and $\Delta^{\mathcal{C}} K^{\bullet\bullet}$ appear to describe excitations between the occupied space and the entire support space defined by $S^{\bullet\bullet}$. However, such an interpretation does not capture changes in the support functions themselves, or any geometric consequences. The similarities between the two formalisms appear to be serendipitous, therefore, and we do not tend to ascribe an interpretation of general changes in the support functions, or consequent density kernel corrections, in terms of single-particle excitations.

## 6.3 Commutator and Gradient Conjugacy in Naphthalene

Improved initial idempotencies imply that the number of iterations spent by the LNVD method initially recovering the purity of the density kernel following NGWF update should be reduced. This is encouraging for the CHRISTOFFEL correction since, in principle, the LNVD method may, in this case, reach a better density kernel in the maximum number of iterations allotted to it, and the next NGWF gradient vector computed will be of a higher quality as a result.

The latter supposition is borne out in Fig. 6.3, which depicts the initial root mean squared value of the matrix elements of the Kohn–Sham commutator after NGWF update, $\hat{\rho}\hat{H} - \hat{H}\hat{\rho}$, and that following density kernel optimisation for the fixed set of NGWFs on that step. Similarly to the trend in the density-matrix occupancy and idempotency, we note that the initial commutator is quite significantly reduced by the introduction of the appropriate geometric correction, $\Delta^{\mathcal{C}} K^{\bullet\bullet}$, while it is actually increased somewhat over the uncorrected case when other corrections involving $\Delta^{\mathcal{P}} K^{\bullet\bullet}$ are used.

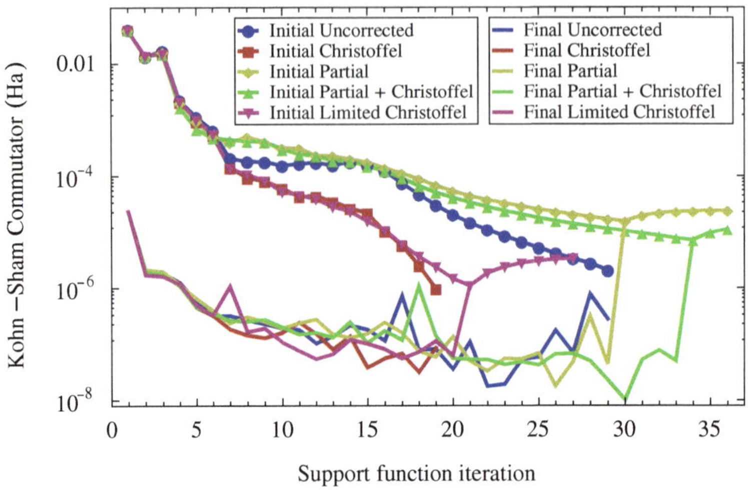


**Fig. 6.3** The root mean squared value of the commutator between the density-matrix and the Kohn–Sham Hamiltonian, computed both immediately after NGWF update and following full density-kernel optimisation for that fixed set of NGWFs, plotted as a function of the NGWF update iteration. Behaviour is shown for the case where no correction is applied to the density kernel upon function change, and when combinations of corrections corresponding to the partial and geometric terms of the absolute derivative are performed

As a result of the very conservative minimum of ten LNVD steps being enforced in these calculations, we observe that all of the methods have ample iterations available to rectify the commutator to a similar level. We note, however, that until they have converged—that is in significantly fewer NGWF steps—the Christoffel corrected calculations (with and without inclusion of the correction in the conjugate gradients trial steps, respectively, CHRISTOFFEL and LIMITED CHRISTOFFEL) generate a lower final commutator at almost all stages. Consequently, if a fewer number of LNVD steps per NGWF step were practicable, as is generally the case as we move to larger systems, we would expect the Christoffel corrected calculations to more appreciably out-perform the alternatives.

An interesting point to note is that when the corrections involving the partial derivatives are used, or if the correction is omitted at the trial steps, we see that the condition known as a "stuck commutator" results. This happens when the NGWFs reach a point where the LNVD method cannot minimise the energy while maintaining the idempotency of the density-matrix, thus providing fruitless attempts at kernel optimisation. The net effect is that the density kernel and NGWFs can only take very small steps which lead toward an irredeemable fixed-point with a poor commutator.

The stuck commutator occurs as a result of inconsistencies between energies and gradients. If the energy minimisation scheme, as it ordinarily stands, is forced to run to very small convergence thresholds, the result is usually a stuck commutator but with an acceptable commutator and idempotency value (this results from numerical noise and the finite spatial extent of the NGWFs). Quite generally, however, if the

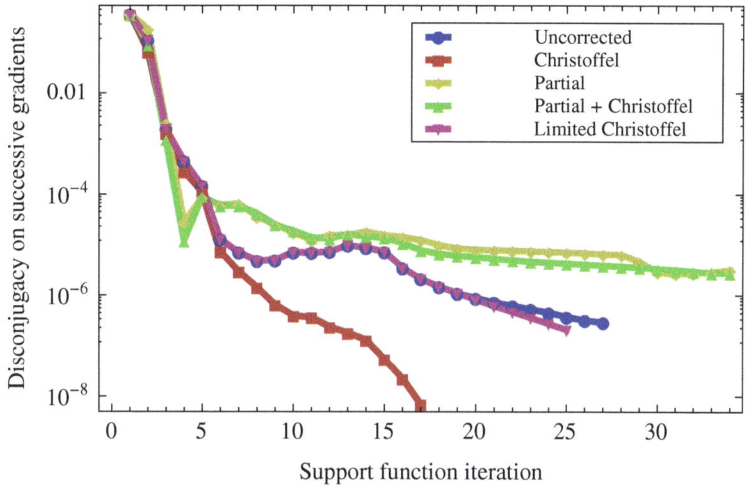


**Fig. 6.4** The absolute value of the overlap integral between a given conjugate gradients search direction and the previous one, as an indicator of the contribution to the extent of failure to meet the conjugacy condition resulting from that iteration, as a function of the NGWF update iteration itself. Behaviour is shown for the case where no correction is applied to the density kernel upon function change, and when combinations of corrections corresponding to the partial and geometric terms of the absolute derivative are performed

total energy appears to be well converged but the commutator remains unacceptably large then the calculation is invalidated since the Kohn–Sham states are then not necessarily orthonormal.

Both the uncorrected energy minimisation scheme and the full Christoffel corrected method had yet to meet a stuck commutator when convergence was attained, however this was not the case for the LIMITED CHRISTOFFEL scheme where corrections are omitted at trial NGWF steps in the Fletcher-Reeves scheme.

As we can see, the calculation is declared to be converged with the CHRISTOFFEL correction after substantially fewer NGWF update iterations than are needed in the alternative approaches. This correction offers a useful saving in computational effort for naphthalene, with a decrease in total computing time of 32% for the set of parameters used. The efficiency could potentially be increased further by taking advantage of the greatly enhanced degree of density kernel purity after NGWF update when the CHRISTOFFEL correction is used, since the effort spent purifying the density kernel using penalty functional methods becomes redundant from an early stage of the calculation onwards and may be obviated.

At least part of the reason behind the more rapid convergence of the energy minimisation algorithm, when the appropriate parallel transport terms are introduced to the density kernel, is evident in Fig. 6.4. This figure shows the deviation from conjugacy between NGWF update steps on successive conjugate gradients iterations, $\langle g^{(i)\alpha} | g_{\alpha}^{(i-1)} \rangle$, which should be identically zero if the conjugate gradients scheme is performing optimally (the current gradient and the sum of all previous gradients should be conjugate). The non-zero value in the disconjugacy between steps is due to the inconsistency between the gradient computed at a point and the energy at which the next gradient is computed.

There are a number of causes of such inconsistency, which we accept because they either bring with them a more rapidly converging method or are difficult to avoid. Firstly, the NGWF update conjugate gradients scheme includes no compensation for the LNVD density kernel update loop which lies (necessarily) within it, changing the energy in a way which is unpredictable to the outer conjugate gradients algorithm. Secondly, the algorithm itself is not optimal when the energy landscape is not perfectly parabolic, and it never is for anything but model systems. Thirdly, the conjugate gradients algorithm is not optimal when any quadratic or higher order curve-fitting is used to calculate the step length and hence to accelerate convergence. Finally, and perhaps not altogether insignificantly, since the sum of previous gradients is itself a rank-one tensor it should, in principle, be parallel transported into the current frame of NGWFs before its inner product is taken with the current gradient. We return to this issue and propose a simple technique to overcome it in Appendix A.3.

Notwithstanding these considerations, Fig. 6.4 strongly suggests that it is of practical benefit to the conjugate gradients minimisation with respect to the NGWFs to geometrically correct the density kernel when the NGWFs are updated; in effect it retrieves some of the inconsistency between energy and gradients in the NGWF optimisation scheme. The reason for this is quite easy to understand since, as far as the optimisation scheme for the NGWFs is concerned, the density kernel is fixed with respect to the frame of NGWFs. However, what we usually fix during NGWF optimisation is the actual elements of the density kernel matrix, the representation of the density-matrix in the frame at which the gradient is computed—which is not quite the same.

The correction introduced by the terms due to the partial derivatives attempt to compensate the density kernel for the change in NGWFs, so that the density matrix operator represented by it remains unchanged to first order. These are not required since we no not assume that the density matrix operator itself is unchanged. The terms due to the Christoffel symbols, however, ensure that the density kernel remains a tensor after NGWF update by accounting for the change in orientation of the frame.

Some of the discrepancy in the NGWF conjugate gradients scheme is removed by the latter correction, leading to a more rapid and stable convergence towards the ground state, while the former is at odds with the assumptions used to compute the derivative of the energy with respect to the NGWFs and so this disrupts the energy minimisation scheme. No significant benefit is yielded, however, as shown by the LIMITED CHRISTOFFEL curve, unless the Christoffel corrections are integrated with the Fletcher-Reeves method and applied when trial NGWF updates are used to find the optimal step length.

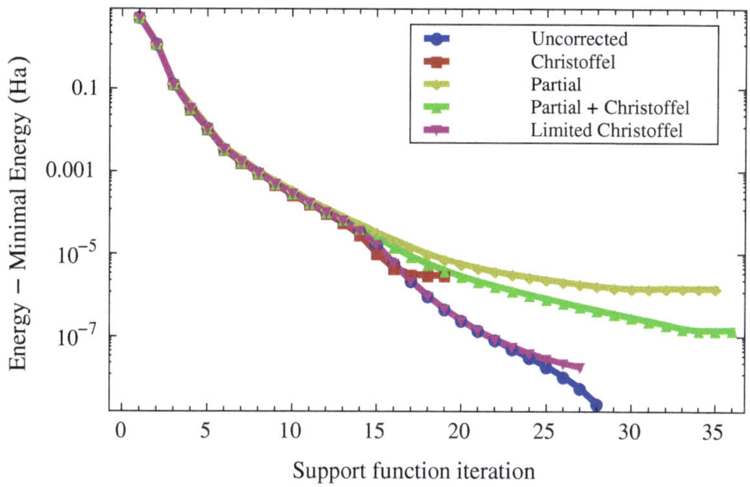


**Fig. 6.5** The total energy as a function of the NGWF update iteration. Behaviour is shown for the case where no correction is applied to the density kernel upon function change, and when combinations of corrections corresponding to the partial and geometric terms of the absolute derivative are performed, and is plotted as the difference from the final total energy given when no correction is used

## 6.4 Total-Energy Convergence in Naphthalene

Figures 6.5 and 6.6 show the convergence of the total energy and the root mean squared value of the NGWF gradient, respectively, evaluated after full density kernel optimisation. The zero of the energy is set to the lowest energy achieved by any of the methods considered, which was that with no correction. All methods behave in a very similar way, by these insensitive measures, until approximately the fifteenth NGWF iteration. Thereafter, the superior update vector conjugacy and commutator in the case of the Christoffel symbol corrected kernel admit a rapid falloff in the magnitude of the gradient vectors and consequently swift convergence. As expected, if the inconsistencies between support function and density kernel optimisation algorithms get smaller as we approach the ground state, the convergence accelerates as they do so.

In the case of the Christoffel correction applied only to the final selected NGWF update step, the LIMITED CHRISTOFFEL method, the energy and gradient convergence is very similar to that in the uncorrected case. This shows that, in spite of improved initial occupancies, idempotencies and commutators after the NGWF update step has been selected, it is crucial to use the correction at all trial steps in the Fletcher-Reeves conjugate-gradients algorithm in order to yield the advantages offered by geometric considerations.

The corrections which involve the partial covariant derivatives of the kernel perform significantly worse than the uncorrected method in terms of convergence speed, for the reasons mentioned above. It is perhaps somewhat disappointing that

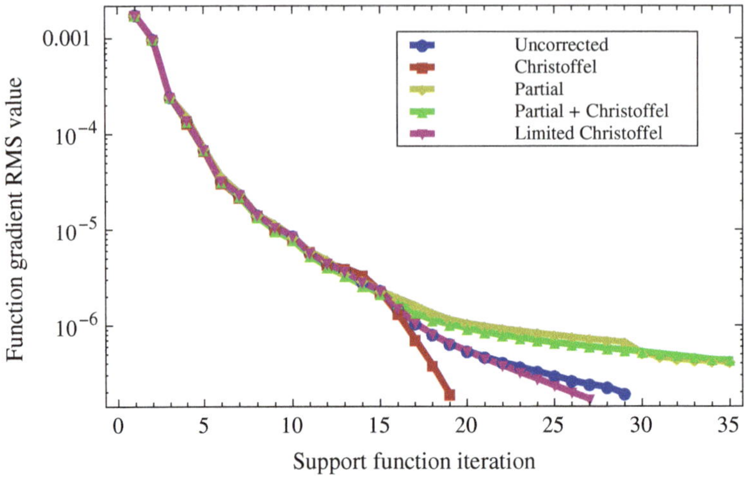


**Fig. 6.6** The root mean squared value of the NGWF gradient, as a function of the NGWF update iteration. Behaviour is shown for the case where no correction is applied to the density kernel upon function change, and when combinations of corrections corresponding to the partial and geometric terms of the absolute derivative are performed

the energy converges to a value of approximately $3 \times 10^{-6}$ Ha (or approximately $5 \times 10^{-6}$ eV/atom) higher in the Christoffel-corrected over the uncorrected case. This effect is somewhat enhanced by the fact that the root mean squared gradient is used in the convergence criterion and not the energy directly, however the qualitative result is not changed. It is, however, rendered redundant by the fact that the error estimate in the final energy, approximated by the root mean squared value of the commutator times the number of NGWFs, is on the order of $1 \times 10^{-5}$ Ha (or approximately $2 \times 10^{-5}$eV/atom) in the uncorrected case and on the order of $4 \times 10^{-6}$ Ha (or approximately $6 \times 10^{-6}$ eV/atom) in the Christoffel corrected case.

## 6.5 Computational Methodology for Oligoacene Polymers

We carried out total energy calculations, using the CHRISTOFFEL corrections to the density kernel, on selected members of the linear oligoacene family of polycyclic aromatic hydrocarbons (the benzene–naphthalene series). Namely, those systems studied, whose calculated highest occupied molecular orbitals are shown in Fig. 6.7 for purposes of illustration, were tetracene ($C_{18}H_{12}$), octacene ($C_{34}H_{20}$), hexadecacene 16-mer ($C_{66}H_{36}$), dotriacontacene 32-mer ($C_{130}H_{68}$), hexapentacontacene 56-mer ($C_{226}H_{116}$) and hexadecahectacene 116-mer ($C_{466}H_{236}$), the lattermost containing 702 atoms in total.

Principally, our goal was to eliminate any possibility that the improvement in the density kernel occupancy and idempotency, the Kohn–Sham commutator

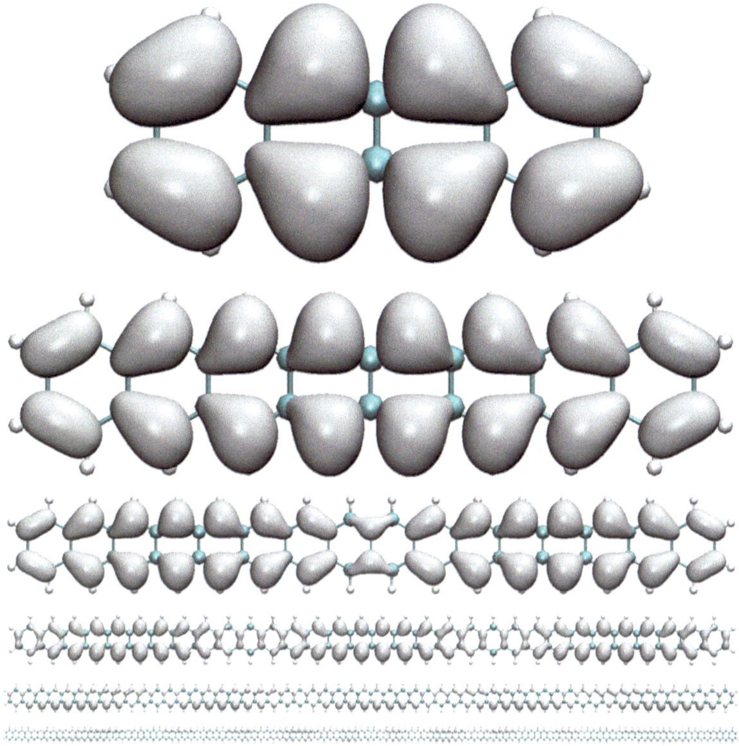

**Fig. 6.7** Highest occupied molecular orbitals of the oligoacene n-mers tetracene ($C_{18}H_{12}$), octacene ($C_{34}H_{20}$), hexadecacene 16-mer ($C_{66}H_{36}$), dotriacontacene 32-mer ($C_{130}H_{68}$), hexapentacontacene 56-mer ($C_{226}H_{116}$) and hexadecahectacene 116-mer ($C_{466}H_{236}$) computed using ONETEP in the PBE approximation

and satisfaction of the gradient conjugacy condition yielded by application of the CHRISTOFFEL correction to naphthalene was an artefact specific to the naphthalene molecule. The correction term arising from the partial derivative of the density kernel contributes to the absolute differential in general, but, as we have shown, it is not appropriate to include it in ONETEP due to the manner in which the energy gradient respect to the support functions is computed in that method; we do not investigate it further in the remainder of this study (Fig. 6.8).

Furthermore, it was of interest to establish what change in the linear-scaling computational pre-factor might be expected to result from such corrections in the case of simple linear hydrocarbons and, more fundamentally, what difference is made to the sensitivity of the energy minimisation algorithm to the degree of the density kernel truncation (applied to admit linear-scaling performance) by these corrections.

The computational methodology for these larger systems was the same as that for naphthalene except that, in order to access the linear-scaling regime with a reasonable computational overhead but still retaining a sufficient accuracy, some of the varia-

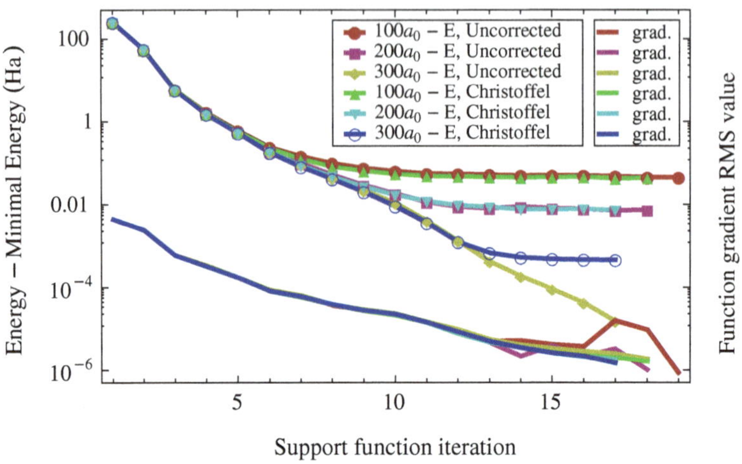


**Fig. 6.8** Convergence of the total energy (uppermost data-points, marked "E") and the root mean squared NGWF gradient (lowermost curves, marked "grad."), with NGWF update steps, for the 116-mer oligoacene molecule. Data is shown for three different values of the density kernel cutoff threshold, both with and without the CHRISTOFFEL correction to the density kernel

tional parameters and convergence criteria were set to more routinely-used values. The threshold on the root mean squared NGWF gradient was reduced to its default value of $2 \times 10^{-6}$ Ha $a_0^{3/2}$ and the number of LNVD steps per NGWF iteration was set to a minimum of 4 and maximum of 8. The kinetic energy cutoff was set to a minimum of 750 eV and the NGWF cutoff radii were set, for carbon and hydrogen, respectively, at 8 $a_0$ and 6.5 $a_0$. The overlap matrix was not inverted exactly but rather using a linear-scaling implementation of Hotelling's algorithm [18].

Atomic positions were generated simply by repetition of the optimised naphthalene structure. It is unlikely that details of the geometry effect the qualitative behaviour of the density kernel purity or total-energy convergence. A large simulation cell was used, of 50 $a_0$ in all directions with additive stretching by multiples of the benzene-ring width in the direction of the polymer chain. The emergence of periodicity in the multiply degenerate highest occupied Kohn–Sham orbital for the larger systems, shown in Fig. 6.7, was most likely due to aliasing effects on the real-space *psinc* grid. We have not investigated the origins of this phenomenon and we concede that it may be an artefact of neglecting the tendency to weak Peierls distortion reported in Ref. [9].

The spatial extent of the density-matrix was probed by performing total-energy minimisation on the largest system with a varying truncation threshold on the elements of the density kernel. A threshold of 300 $a_0$ was sufficient to provide a 100% matrix filling. We found that a large density kernel cutoff was required for these systems, due to a rapid (approximately proportional to the inverse polymer length) decrease in the HOMO-LUMO gap with system size, from 1.45 eV in the 4-mer down to an insubstantial 0.02 eV from the 32-mer onwards. The convergence

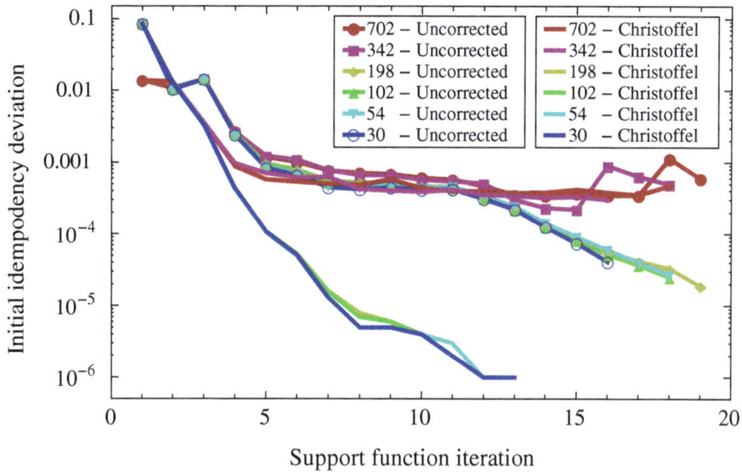


**Fig. 6.9** The density-matrix idempotency error computed immediately after an NGWF update has been applied, with a 100 $a_0$ density kernel truncation threshold, plotted against NGWF iteration number for oligoacene polymers, both with and without the CHRISTOFFEL correction. The polymers are labelled by their total number of atoms

of the total-energy and the root mean squared NGWF gradient under these circumstances, with and without CHRISTOFFEL corrections to the density kernel, is depicted in Fig. 6.9.

A difference in total energies of approximately $5 \times 10^{-5}$ Ha/atom (approximately $1 \times 10^{-3}$ eV/atom) was found between calculations with a density kernel cutoff of 100 $a_0$ (35% matrix filling) and 200 $a_0$ (69% matrix filling). It was not possible to converge the energy reliably at a density kernel cutoff of 50 $a_0$ (18% matrix filling) and both the Kohn–Sham commutator and idempotency error remained unacceptably large in this case.

We concluded that a threshold of 100 $a_0$ was thus ideal for our purposes, providing a reasonable tolerance for the energy but being quite definitely near the minimum of what is required for stable convergence in the largest system. Thus, it provided the sensitivity under perturbations to the density-matrix non-locality to allow us to investigate whether the geometric corrections aid or destabilise convergence in a marginal case.

We found that fewer NGWF iterations were needed to converge the calculations when the geometric corrections were applied with kernel truncation, though negligibly so. The corrections do influence the convergence behaviour and final total-energy somewhat, tending to slightly reduce gradients values, however the largest total-energy difference introduced, at $1 \times 10^{-6}$ Ha/atom in the 100 $a_0$ case, was within the accuracy of the method.

## 6.6 Geometric Density Kernel Corrections in Oligoacenes

The preservation of the idempotency of the density-matrix to first order in the NGWF update vector is the most interesting and promising aspect of the CHRISTOFFEL form of geometric correction. It was crucial, therefore, to investigate the effects of system size and density kernel truncation on this property.

The behaviour of the idempotency error for the oligoacene polymers is shown in Fig. 6.9, where a 100 $a_0$ density kernel truncation threshold was used. The density kernel was fully filled for systems up to and including the 32-mer (198 atom) polymer, while for the 56-mer (342 atom) and 116-mer (702 atom) it stood at 68% and 35%, respectively. Occasional occupancy error correction, a non-linear scaling step, was needed to ensure good convergence in the lattermost case.

It is immediately apparent from this plot that the substantial reduction provided by the geometric corrections both in the initial idempotency error, at approximately two orders of magnitude, and the number of NGWF iterations required for convergence, is remarkably system-independent for those systems with an non-truncated density kernel. This amelioration suggests that the corrections may be generally beneficial in systems with no density kernel truncation, and also may hint at some underlying system-independent limitations in the method in terms of how rapidly the idempotency error upon NGWF update can be removed (approximately exponentially, we hasten to note, when geometric corrections are employed).

For systems with non-trivial truncation, unfortunately, the small numerical value of the corrections in practice appears to be overwhelmed by noise in the density kernel. The corrected calculations tend to possess a purer density-matrix up to approximately the twelfth NGWF iteration, and do so at the calculation's conclusion, but the uncorrected calculations may persevere at intermediate stages. Thus, the density kernel corrections do not provide any additional robustness against heavy density kernel truncation, perhaps unremarkably since the corrections themselves are computed via products of truncated matrices.

## 6.7 Commutator and Conjugacy Condition in Oligoacenes

The commutator between the density kernel and the Hamiltonian, computed after the NGWFs are renewed, exhibits similar trends to the idempotency. An appreciable and valuable reduction is provided by the geometric corrections for systems in which the density kernel is non-truncated but no systematic change can be attributed where truncation does take place. For truncation at longer length scales than the 100 $a_0$ used in this study, we would expect to observe an intermediate behaviour between the two regimes. In any case, in spite of the Hamiltonian being fully recomputed, the CHRISTOFFEL correction at best substantially improves, at worst leaves unchanged, the accuracy bound on the ground-state energy provided by the commutator (Fig. 6.10).

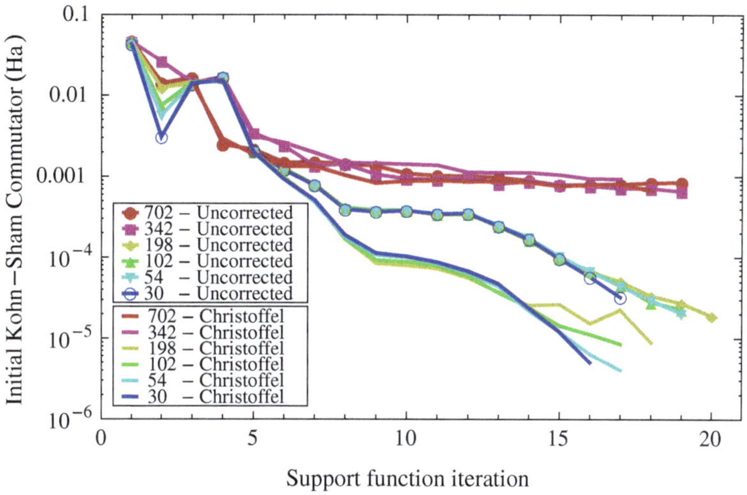


**Fig. 6.10** The root mean squared value of the Kohn–Sham commutator computed immediately after an NGWF update has been applied, with a 100 $a_0$ density kernel truncation threshold, plotted against NGWF iteration number for oligoacene polymers, both with and without the CHRISTOFFEL correction. The polymers are labelled by their total number of atoms

The deviation from conjugacy between successive NGWF gradients provides an approximate indicator of the performance of the conjugate gradients algorithm used to minimise the total-energy. This quantity, divided by the number of support functions in order to provide a system-size independent measure, is depicted for the polymers studied in Fig. 6.11.

For the four smallest systems, we conclude that the geometric corrections substantially aid the algorithm and improves the quality of the NGWF steps by up to two orders of magnitude by this measure. Interestingly, for the two larger systems where density kernel truncation incapacitates the geometric corrections so far as they may reduce the idempotency deviation and the commutator, those corrections nonetheless result in noticeable reduction of the gradient disconjugacy, particularly for the largest system where some occupancy corrections were applied.

The reduction of the conjugacy error is attributable to an improved agreement between energy changes predicted using the gradient vectors and the actual energies arrived at. This accords with our analytical prediction that without the geometric corrections entering the density kernel used to compute the latter energies, there is a inconsistency between gradients and energies regardless of whether density kernel truncation is applied. This inconsistency may result in poorer NGWF step lengths being used and so a conjugate gradients method which is further from optimal for the energy minimisation problem.

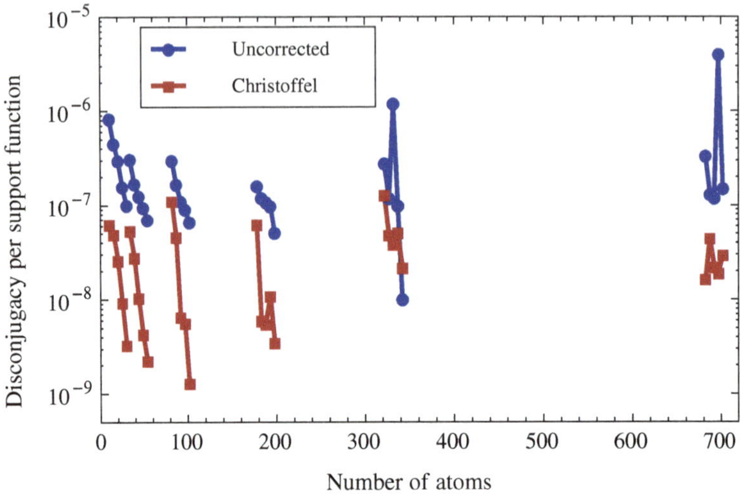


**Fig. 6.11** The conjugacy error between successive NGWF gradients, both with and without geometric corrections, in total-energy calculations on selected oligoacene polymers. The disconjugacy is divided by the number of NGWFs in order to provide a system-size independent measure of the performance of the conjugate gradients algorithm. The value at the final five NGWF iterations of each calculation is shown, which a slight artificial displacement in the horizontal axis for clarity; the true number of atoms is given by the horizontal position of the final data-point of each set

## 6.8 Computational Performance in Oligoacenes

Linear polymers such as the oligoacenes we have studied are the ideal candidates for a linear-scaling method such as ONETEP because the NGWF overlap and Hamiltonian matrices rapidly become very sparse with increasing polymer length, facilitating optimal parallelisation of computational effort. Moreover, since the non-locality of the density-matrix occurs predominantly in one dimension, the error due to density kernel truncation is easily analysed for such systems.

The density kernel cutoff threshold used in our study, which was acceptable but not generous, allowed for the linear-scaling regime to be easily accessed for the set of polymers we have studied, see Fig. 6.12. On 28 cores of a contemporary commodity supercomputer we obtained a pre-factor of approximately 1.12 min/atom for uncorrected calculations which reduced to approximately 1.00 min/atom when geometric corrections were applied.

This increase in performance seems rather modest in light of the improvements furnished by the geometric corrections to the density-matrix idempotency, Kohn–Sham commutator and the gradient conjugacy condition. It comes primarily as a consequence of slightly more NGWF iterations being required for convergence in the uncorrected case, also shown in Fig. 6.12, but also the tendency for the uncorrected calculations to use more than the minimum number of LNV iterations per NGWF step more frequently than in the corrected case. These are two aspects of the same

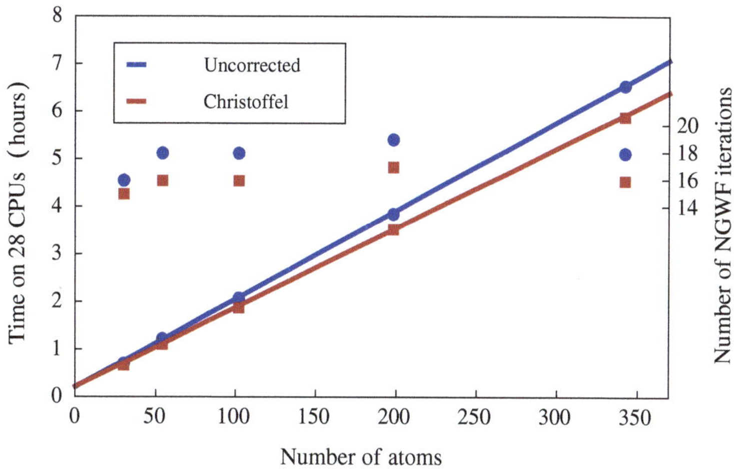


**Fig. 6.12** Time taken for a converged total-energy minimisation (left axis, lines) and the number of NGWF iterations required to do so (right axis, solitary points) for selected oligoacene polymers, both with and without geometric density kernel corrections for NGWF updates

phenomenon; the Kohn–Sham commutator and NGWF gradients are improved by the geometric corrections.

There are a number of ways in which the analysis favours the uncorrected case in terms of overall computational expense and makes a direct comparison difficult. This is unsurprising, of course, since the ONETEP method has been developed and optimised without the presence of geometrically corrective terms.

Firstly, upper limits are placed on the density kernel and NGWF step lengths in order to prevent excessive, destructive steps being taken as a result of impure density-matrices. These limits are necessary for stable convergence in general, but are probably excessively severe for the corrected case where the density-matrix is generally of a much better quality. In order to perform a balanced comparison of methods in this regard it would be necessary to allow the maximum step lengths to be dynamically computed on the basis of the idempotency error or commutator, and this is perhaps an interesting avenue for future investigation.

Additionally, checks and corrections for density-matrix idempotency deviation, which are performed when either the NGWFs of density kernel are changed, are necessary in the uncorrected case but are wasteful, and perhaps counter-productive, in the latter stages of a geometrically corrected calculation.

Most significantly, however, a minimum of four LNVD iterations per NGWF step were used to provide stable convergence of the energy in both uncorrected and corrected calculations and thus to perform a comparison. In practice, particularly in systems with no density kernel truncation, this was most likely not necessary for the corrected calculations at all stages. For small systems, LNVD optimisation may not be required at every NGWF step with a geometrically corrected density kernel.

Finally, and aside from issues of convergence behaviour, is the issue of the convergence criterion used to halt the calculation. The root mean squared value of the NGWF gradient is used as the single criterion in the ONETEP code at present. This is a matter of pragmatism since it gives a good indication of whether the calculation, if not converged in terms of the total-energy, has stopped making improvements to the Kohn–Sham eigenfunctions. In practice, of course, we use this measure in conjunction with the behaviour of the total energy in the final number of NGWF steps to judge whether true convergence has been obtained.

We are of the view that both of these two criteria and the Kohn–Sham commutator should be used together to provide a threshold of convergence, since the latter also provides an estimator of the error (due to non-orthogonality between Kohn–Sham eigenstates) in the total-energy. Were the Kohn–Sham commutator used as a convergence criterion, in its capacity as an energy uncertainty estimate, instead of the NGWF gradient, it appears that the speedup due to geometric corrections would be judged to be somewhat greater.

## 6.9 Concluding Remarks

In conclusion, we have found that the corrections to the density kernel for NGWF update furnished by the Christoffel terms provide a valuable enhancement to the stability and rapidity of total-energy convergence in selected oligoacene polymers. In particular, the analytical prediction that the purity of the density-matrix should be preserved to first order is borne out in practice, and additionally there is a significant improvement to the Kohn–Sham commutator computed after NGWF update. The partial derivative of the density kernel with respect to NGWFs provides terms which are not appropriate to the ONETEP method. When the geometric corrections are included in the Fletcher-Reeves conjugate gradients scheme, the result is a dramatic reduction in the conjugacy error between successive NGWF gradients. We expect that such corrections may render a greater variety of realistic, problematic systems routinely amenable to linear-scaling ab initio methods.

## References

1. C.-K. Skylaris, P.D. Haynes, A.A. Mostofi, M.C. Payne, Introducing ONETEP: Linear-scaling density functional simulations on parallel computers. J. Chem. Phys. **122**, 084119 (2005)
2. P.D. Haynes, C.-K. Skylaris, A.A. Mostofi, M.C. Payne, Elimination of basis set superposition error in linear-scaling density-functional calculations with local orbitals optimised in situ. Chem. Phys. Lett. **422**, 345 (2006)
3. F. Mauri, G. Galli, R. Car, Orbital formulation for electronic-structure calculations with linear system-size scaling. Phys. Rev. B **47**(15), 9973 (1993)
4. F. Mauri, G. Galli, Electronic-structure calculations and molecular-dynamics simulations with linear system-size scaling. Phys. Rev. B **50**(7), 4316 (1994)

5. J. Kim, F. Mauri, G. Galli, Total-energy global optimizations using nonorthogonal localized orbitals. Phys. Rev. B **52**(3), 1640 (1995)
6. P. Ordejón, D.A. Drabold, R.M. Martin, M.P. Grumbach, Linear system-size scaling methods for electronic-structure calculations. Phys. Rev. B **51**(3), 1456 (1995)
7. C.-K. Skylaris, A.A. Mostofi, P.D. Haynes, O. Diéguez, M.C. Payne, Nonorthogonal generalized Wannier function pseudopotential plane-wave method. Phys. Rev. B **66**(3), 035119 (2002)
8. G. Malloci, G. Mulas, G. Cappellini, C. Joblin, Time-dependent density functional study of the electronic spectra of oligoacenes in the charge states -1, 0, +1, and +2. Chem. Phys. **34**(1-3), 43 (2007)
9. T.A. Niehaus, M. Rohlfing, F. Della Sala, A. Di Carlo, T. Frauenheim, Quasiparticle energies for large molecules: a tight-binding-based Green's-function approach. Phys. Rev. A **71**(2), 022508 (2005)
10. P. Elliott, F. Furche,K. Burke. Excited states from time-dependent density functional theory. Reviews in Computational Chemistry (Wiley, 2009), pp. 91–165
11. A.A. Mostofi, P.D. Haynes, C.-K. Skylaris, M.C. Payne, Preconditioned interative minimisation for linear-scaling electronic structure calculations. J. Chem. Phys. **119**, 8842 (2003)
12. D. Baye, P.-H. Heenen, Generalised meshes for quantum mechanical problems. J. Phys. A: Math. Gen. **19**, 2041 (1986)
13. J.P. Perdew, K. Burke, M. Ernzerhof, Generalized Gradient Approximation Made Simple. Phys. Rev. Lett. **77**(18), 3865 (1996)
14. X.-P. Li, R.W. Nunes, D. Vanderbilt, Density-matrix electronic-structure method with linear system-size scaling. Phys. Rev. B. **47**(16), 10891 (1993)
15. R.W. Nunes, D. Vanderbilt, Generalization of the density-matrix method to a nonorthogonal basis. Phys. Rev. B **50**(23), 17611 (1994)
16. M.S. Daw, Model for energetics of solids based on the density matrix. Phys. Rev. B **47**(16), 10895 (1993)
17. R. McWeeny, Some recent advances in density matrix theory. Rev. Mod. Phys. **32**(2), 335 (1960)
18. T. Ozaki, Efficient recursion method for inverting an overlap matrix. Phys. Rev. B **64**(19), 195110 (2001)

# Chapter 7
# Tensorial Aspects of Calculating Hubbard $U$ Interaction Parameters

First-principles Hubbard U parameters for use in ab initio methods for strongly-correlated systems, such as DFT + $U$ and DFT+DMFT, have posed a challenge to theoretical calculations for some considerable time. These parameters are the matrix elements of the screened Coulomb operator, $\hat{U}^{(\sigma)}\left(\mathbf{r}, \mathbf{r}'\right)$, which describes the effective interactions between single-particle states (either orthonormal Kohn–Sham eigenstates or quasiparticles) which form the correlated subspaces.

One might ask which screening mechanisms should be included in $\hat{U}$ for use with a given method. If we wish to provide a Hubbard $U$ parameter, or tensor, for the DFT + $U$ or DFT+DMFT methods, then the screening mechanisms associated with the correlated subspaces should not be treated on the same footing as those of the remainder of the system. For such methods, and we concentrate on DFT + $U$ here, we must incorporate the nature of the correlated subspaces themselves in the construction of the screened interaction operator, not just in its matrix elements, embedding an effective low-energy Hubbard model in the Kohn–Sham system using a technique known as "down-folding".

For specific details of some of the methods available for computing interaction parameters for various purposes, we refer the reader to the original sources for the self-consistent linear density-response [1–3], constrained Random Phase Approximation (cRPA)[4, 5], constrained LDA [6–8] and constrained adiabatic LDA [9] methods.

Here we focus, in particular, on the first two such methods, which are constructed to provide parameters appropriate to the DFT + $U$ method. We generalise these methods to the formalism of both a nonorthogonal representation of single-particle states and nonorthogonal Hubbard projectors, mentioning any obstacles to linear-scaling performance where appropriate. The expressions in this chapter have not been implemented in the ONETEP code as yet, but they may serve as a guide for incorporation of first-principles techniques for the Hubbard $U$ parameter into a linear-scaling method. Finally, we provide the necessary expressions for the parallel transport of these interaction tensors after they have been calculated. This strategy may prove to partially obviate the re-computation of $U$ at each step of the projector self-consistency procedure of Chap. 3 and Ref. [10], admitting very efficient

D. D. O'Regan, *Optimised Projections for the Ab Initio Simulation of Large and Strongly Correlated Systems*, Springer Theses, DOI: 10.1007/978-3-642-23238-1_7,

implementation of a proposed combination of first-principles schemes for both Hubbard projectors and parameters.

## 7.1 The Linear Density-Response Method

We first consider the generalisation of the self-consistent linear density response approach, described in Refs. [2, 3], to accommodate non-orthogonal Hubbard projectors. This method is, itself, a development of a previously proposed technique described in Ref. [1], giving an enhanced description of screening mechanisms by allowing for the self-consistent reorganisation of the orbitals in response to perturbations.

It has been shown, for the exact exchange-correlation functional, that the total-energy of an open atomic system in contact with a reservoir varies in a piecewise-linear manner with atomic orbital occupancies (technically we may consider time-averaging the system in an ensemble state composed of a fluctuating linear-combination of pure integer-occupancy states) but that for LDA-type exchange-correlation functionals this does not generally hold [11].

This is generally due to lack of a derivative discontinuity with respect to particle number in such functionals, a pathology which is responsible for a significant fraction of their characteristic underestimation of the insulating gap. We refer the reader to Ref. [12] for an overview of the deficiencies of such functionals in this regard.

In practice, the LDA exhibits smooth, concave (approximately parabolic) energy-occupation curves, and the minimum of such curves generally occur at fractional values for the total number of electrons. The problem is particularly pronounced in strongly-correlated systems, where the LDA may lower the energy by fractionally occupying the degenerate, spatially localised, states which should be split into occupied and virtual bands. A confounding issue arising from partial occupation of the localised states is that the spurious Coulomb repulsion between them usually causes some reduction of their spatial localisation and this, in turn, causes an underestimation of local magnetic moments and a further collapse of the insulating gap due to diminished exchange splitting.

In the linear-response method, the DFT + $U$ penalty functional is interpreted as a correction for the spurious curvature of the total-energy of the system, with respect to the occupancies of the localised orbitals away from integer value, which is exhibited by approximate exchange-correlation functionals. The Hubbard $U$ is identified as the erroneous curvature to be cancelled by the DFT + $U$ correction, and the correlated subspaces, which may deviate from spherical symmetry due to their perturbing environment and which measure the curvature and to which the correction is applied, are those spanned by the Hubbard projectors.

Denoting the total occupancy of site $I$ with spin $\sigma$ by $N^{(I)(\sigma)}$, a scalar Hubbard $U$ parameter may be computed as

$$U^{(I)(\sigma)} = \frac{d^2 E}{d\left(N^{(I)(\sigma)}\right)^2} - \frac{d^2 E_{independent}}{d\left(N^{(I)(\sigma)}\right)^2}, \tag{7.1}$$

where, as detailed in Refs. [1, 2], the independent-particle (kinetic) contribution to the density-response is deducted in order to leave only the purely interacting (due to Hartree and exchange-correlation terms and comparatively small in magnitude) components appropriate to the down-folded Hubbard model.

The linear density-response method, by construction, measures the extent to which the description of the Coulomb interactions, represented by the $U$ parameter, are underestimated in the ground-state response function. It is, of course, very desirable that the Hubbard $U$ should be calculated using the response of the DFT + $U$ ground state, and not that of the uncorrected DFT ground state. A promising self-consistency scheme for this method was proposed in reference [3], whereby the response, $U^{out}$, is calculated at a number of input values, $U^{in}$, after which the extrapolation to $U^{out}\left(U^{in}=0\right)$ is deemed to be the self-consistent value if a good fit can be obtained.

### 7.1.1 Towards a Projector-Decomposed Method

Let us investigate the generalisation of the linear-response method to compute orbital-resolved $U$ tensors. It has been shown, for example in Ref. [1], that components of the $U$ tensor may be rather strongly dependent on the symmetry of the localised orbitals, for example those of $3d$ orbitals of $e_g$ and $t_{2g}$ symmetry, and so an averaged scalar response may not be realistic in such cases.

A rank-2 $U$ is associated with the curvature of the total energy with respect to two different orbital densities, whereas in the rank-4 case it is the curvature with respect to two different subspace density matrices. We will assume that interactions are diagonal in the spin-index, although the generalisation to off-diagonal interactions between spins is straightforward in principle.

To compute the Hubbard $U$ within linear density-response theory, we apply perturbing potentials to the correlated subspaces in order to numerically measure the linear-response at the ground state. The perturbing potential, for each site, most generally takes the form of a Hermitian operator with orbital-dependent matrix elements, as in the tensor form

$$\hat{\alpha}^{(I)}=|\varphi^{(I)m}\rangle\alpha_m^{(I)m'}\langle\varphi_{m'}^{(I)}|=|\varphi_{m''}^{(I)}\rangle O^{(I)m''m}\alpha_m^{(I)m'}\langle\varphi_{m'}^{(I)}|, \tag{7.2}$$

where here and in the remainder of this section we may drop the spin index and consider only the computation of $U$ for one site at a time, for simplicity.

We seek the curvature of the total-energy with respect to subspace occupancy matrices, and of course this task is simplified if the derivative may be reformulated in terms of the matrix elements $\alpha_m^{(I)m'}$ of the perturbation instead. As suggested by Ref. [2], we may move from an expression of the ground-state energy for each value of the subspace perturbation tensor,

$$E\left[\alpha_m^{(I)m'}\right]=\min_{n(\mathbf{r})}\left\{E\left[n\left(\mathbf{r}\right)\right]+\alpha_m^{(I)m'}n_{m'}^{(I)m}\right\}, \tag{7.3}$$

to a representation of these energies in terms of the subspace occupancy tensor, via a Legendre transform.

Specifically, if the matrix elements of the subspace density-matrix are chosen as the independent variables, the perturbation-dependent energy may be re-expressed as

$$E\left[n_m^{(I)m'}\right] = \min_{\alpha_{m'}^{(I)m}} \left\{E\left[\alpha_{m'}^{(I)m}\right] - n_m^{(I)m'} \alpha_{m'}^{(I)m}\right\}. \tag{7.4}$$

We may now take the second derivative of this energy functional, for an arbitrary pair of projector indices, $m^4, m^5$, separately with respect to two copies of the subspace occupancy tensor, whereupon we find that

$$\begin{aligned} \frac{d^2 E\left[n_{m^4}^{(I)m^5}\right]}{dn_{m''}^{(I)m'''} dn_{m'}^{(I)m}} &= \frac{d}{dn_{m''}^{(I)m'''}} \left[-\delta_{m^4}^{m'} \delta_m^{m^5} \alpha_{m^5}^{(I)m^4} \left(n_{m^4}^{(I)m^5}\right)\right] \\ &= -\frac{d\left[\alpha_m^{(I)m'} \left(n_{m'}^{(I)m}\right)\right]}{dn_{m''}^{(I)m'''}} \\ &\equiv \left(-\chi^{-1}\right)^{(I)m''m'}_{\ \ \ m \ \ \ \ m'''}. \end{aligned} \tag{7.5}$$

Thus, in a manner reminiscent of Janak's theorem [13] applied to each correlated subspace, we find that the second derivative of the energy with respect to the subspace occupancy tensor is equivalent to the negative of the inverse of the first derivative of the occupancy tensor with respect to the probe potential tensor, namely that of the static, irreducible linear density-response function $\hat{\chi}$.

The irreducible linear-response function is related to its independent particle contribution $\hat{\chi}_0$ and the screened Coulomb interaction operator $\hat{W}$. The latter may be thought of as built from the bare Coulomb operator, the dielectric function and, if dynamical screening is included (as it is in TDDFT [14]), the second derivative of the exchange-correlation energy known as the exchange-correlation kernel $f_{xc}$). By definition, the Hubbard $U$ parameter is formed from the matrix elements of $\hat{W}$ in the basis of Hubbard projectors.

The total and independent-particle response functions are related Dyson equation [15] given by

$$\hat{\chi} = \hat{\chi}_0 + \hat{\chi}_0 \hat{W} \hat{\chi} \tag{7.6}$$

and we may express the total function as a sum of the independent-particle and an interacting remainder, since

$$\begin{aligned} \left(\hat{1} - \hat{\chi}_0 \hat{W}\right) \hat{\chi} &= \hat{\chi}_0 \\ \Rightarrow \hat{\chi} &= \left(\hat{1} - \hat{\chi}_0 \hat{W}\right)^{-1} \hat{\chi}_0 \end{aligned}$$

$$
\begin{aligned}
&\Rightarrow \hat{\chi}^{-1} = \hat{\chi}_0^{-1}\left(\hat{1} - \hat{\chi}_0 \hat{W}\right) \\
&\Rightarrow \hat{\chi}^{-1} = \hat{\chi}_0^{-1} - \hat{W} \\
&\Rightarrow \hat{W} = \hat{\chi}_0^{-1} - \hat{\chi}^{-1},
\end{aligned} \tag{7.7}
$$

providing a concrete justification of Eq. 7.1.

The full and independent-particle inverse response tensors, which have the same units as Coulomb interactions, are evaluated by numerical finite differences, according to Eq. 7.5. The independent-particle, or non-interacting, component is given by the response calculated without updating the Hamiltonian for changes to the ground-state density. These are then used to compute the Hubbard $U$ parameter using Eq. 7.7, approximations and properties of which we go on to describe in the following subsections.

### 7.1.2 The Non-Locally Resolved Four-Index U Tensor

It is possible to use one of a hierarchy of approximations for the linear-response function in practice, depending on whether we wish to include the fully non-local (within the extent of the subspace) density response, a scalar response in which the orbital degrees of freedom of both the perturbation and the probe are fully averaged (as is most commonly used), or an intermediate level of approximation in which one spatial degree of freedom is integrated over and one is retained.

Suppressing the removal of the independent-particle response function so that $\hat{W} = -\hat{\chi}^{-1}$, for brevity of nation, the subspace-projected linear response function is most generally given by the rank-4 tensor

$$
\begin{aligned}
\chi^{(I)m''m'}_{\ \ m \quad m'''} &= \langle \varphi^{(I)}_m \varphi^{(I)m''} | \hat{\chi}\left(\mathbf{r}, \mathbf{r}'\right) | \varphi^{(I)m'} \varphi^{(I)}_{m'''} \rangle \\
&= \frac{dn^{(I)m'}_m}{d\alpha^{(I)m'''}_{m''}} = \frac{dn^{(I)}_{mm^4}}{d\alpha^{(I)}_{m''m^5}} O^{(I)m^4 m'} O^{(I)}_{m^5 m'''},
\end{aligned} \tag{7.8}
$$

that is the negative of the inverse of Eq. 7.5. This describes the individual response of the components of the subspace occupancy matrix with respect to individual changes to the perturbation matrix.

In this formalism, the perturbing potential at location $\mathbf{r}'$, applied via the projectors labelled $m''$ and $m'''$, induce a change to the density at $\mathbf{r}$, via the response function $\hat{\chi}$, which is in turn probed by the projectors labelled $m$ and $m'$. It may be readily observed that the tensor

$$
-\chi^{(I)m''\,m'''}_{\ \ m \ \ m'} = -\chi^{(I)m''m^4}_{\ \ m \quad m^5} O^{(I)}_{m^4 m'} O^{(I)m^5 m'''} \tag{7.9}
$$

is symmetric since the response operator is Hermitian,

$$
\chi^{(I)m''\,m'''}_{\ \ m \ \ m'} = \chi^{(I)m'''m''}_{\ \ m' \ m}. \tag{7.10}
$$

The projector-dependent $U$ tensor is then evaluated in the convenient form given by

$$U^{(I)\ m''m'''}_{mm'} = O^{(I)}_{m'm^4} \left(-\chi^{-1}\right)^{(I)m^4\ m'''}_{m\ \ m^5} O^{(I)m^5m''}. \tag{7.11}$$

## 7.1.3 The Scalar Interaction U

The most commonly used approximation for the screened Coulomb interaction, at the time of writing, is where the both the perturbing projectors and probing projectors are contracted over before use, indeed before their computation, providing a scalar density-density Coulomb interaction $U$.

The response function for each site is given in terms of the ingredients of the more general four-index formalism by

$$\chi^{(I)} = \chi^{(I)m'm}_{m\ \ m'} = \frac{dn^{(I)m}_{m}}{d\alpha^{(I)m'}_{m'}}, \quad U^{(I)} = -\chi^{(I)-1} \tag{7.12}$$

which may be conveniently evaluated since the perturbation is applied over the entire subspace instead of individually to the projectors.

In this approximation, where we consider only a scalar occupancy responding to a scalar perturbation for each site, the perturbation at $\mathbf{r}'$ is that applied by all projectors on the site in unison and the response at $\mathbf{r}$ is probed together by all the projectors. The usual DFT + $U$ penalty functional in this fully averaged approximation is thus given in tensorially-invariant form by the expression

$$\sum_{I\sigma} \frac{1}{2} U^{(I)\ mm'}_{mm'} \left[ n_{m''}^{\ m''} - n_{m''}^{\ m'''} n_{m'''}^{\ m''} \right]^{(I)(\sigma)}. \tag{7.13}$$

## 7.1.4 The Locally Resolved Two-Index U Tensor

It is of interest to investigate the intermediate approximation, that is where one orbital degree of freedom is retained in the $U$ tensor and the other is integrated over. We are faced with two possible singly projector-decomposed, tensorially invariant, DFT + $U$ models of the penalty-functional form.

These, namely, correspond to that formed by considering changes in each element of the occupancy matrix while using a uniform perturbation (averaging the linear response over all perturbation channels), that is

$$\sum_{I\sigma} \frac{1}{2} U^{(I)\ m'm''}_{mm''} \left[ n_{m'}^{\ m} - n_{m'}^{\ m'''} n_{m'''}^{\ m} \right]^{(I)(\sigma)}, \tag{7.14}$$

or that formed by taking the response of the total occupancy of the subspace to individual orbital-dependent perturbations (averaging the linear response over all probe channels), that is given by

$$\sum_{I\sigma} \frac{1}{2} U^{(I)\ m''m'}_{m''m} \left[ n_{m'}{}^{m} - n_{m'}{}^{m'''} n_{m'''}{}^{m} \right]^{(I)(\sigma)} . \tag{7.15}$$

Fortunately, this ambiguity is fully overcome by noting that $U$ tensors of the form of Eq. 4.25 (having a symmetric screened Coulomb operator, as we expect for the static limit where $U$ is real-valued, possess the symmetry property

$$U^{(I)\ m''m'''}_{mm'} = U^{(I)\ m'''m''}_{m'm} . \tag{7.16}$$

In order to see this more clearly, perhaps, consider the density response operators (recalling that $\hat{\chi}\left(\mathbf{r}, \mathbf{r}'\right) = \hat{\chi}\left(\mathbf{r}', \mathbf{r}\right)$), where we see that, as required,

$$\begin{aligned} \frac{dn^{(I)m''}_{m''}}{d\alpha^{(I)m}_{m'}} &= \langle \varphi^{(I)}_{m''} \varphi^{(I)m'} | \hat{\chi}\left(\mathbf{r}, \mathbf{r}'\right) | \varphi^{(I)m''} \varphi^{(I)}_{m} \rangle \\ &= \langle \varphi^{(I)m''} \varphi^{(I)}_{m} | \hat{\chi}\left(\mathbf{r}', \mathbf{r}\right) | \varphi^{(I)}_{m''} \varphi^{(I)m'} \rangle \\ &= \langle \varphi^{(I)}_{m} \varphi^{(I)m''} | \hat{\chi}\left(\mathbf{r}, \mathbf{r}'\right) | \varphi^{(I)m'} \varphi^{(I)}_{m''} \rangle \\ &= \frac{dn^{(I)m'}_{m}}{d\alpha^{(I)m''}_{m''}} = \chi^{(I)m''m'}_{m\ \ \ m''} = \chi^{(I)m'}_{m} . \end{aligned} \tag{7.17}$$

Both the perturbation-averaged and probe-averaged DFT + $U$ functionals, therefore, give precisely the same two-index response function and hence the same two-index $U$ tensor

$$U^{(I)m'}_{m} \equiv U^{(I)\ m'm''}_{mm''} = \left( -\chi^{(I)-1} \right)_{m}^{\ m'} \tag{7.18}$$

From a computational point of view, the former formulation is much simpler to carry out in practice, since in this way a single scalar perturbation can be applied to each site, as it usually is in the linear-response approach, in order to calculate the projector-decomposed screened response of all elements of the occupancy tensor.

### 7.1.5 Generalisation of the DFT + U Potential and Ionic Forces to the Tensorial Formalism

The DFT + $U$ term in the Kohn–Sham potential, generally given (for real valued $U$ tensors) by

$$\hat{V}^{(\sigma)}_{DFT+U} = \frac{\partial E_{DFT+U}}{\partial \hat{\rho}^{(\sigma)}} = \sum_I |\varphi^{(I)m}\rangle V^{(I)(\sigma)m'}_{m} \langle \varphi^{(I)}_{m'}|, \tag{7.19}$$

has matrix elements in the fully orbital-dependent case given by

$$\begin{aligned} V^{(I)(\sigma)m'}_{m} &= O^{(I)}_{mm''} \frac{\partial E_{DFT+U}}{\partial n^{(I)(\sigma)m'''}_{m''}} O^{(I)m'''m'} \\ &= \frac{1}{2} U^{(I)\,m'm'''}_{mm''} \left[ \delta^m_{m'} \delta^{m''}_{m'''} - n_{m'}^{\ m} \delta^{m''}_{m'''} - \delta^m_{m'} n_{m'''}^{\ m''} \right]^{(I)(\sigma)} \\ &= \frac{1}{2} U^{(I)\,m'm'''}_{mm''} \left[ \delta^m_{m'} \delta^{m''}_{m'''} - 2 n^{(I)(\sigma)m}_{m'} \delta^{m''}_{m'''} \right] \\ &= \frac{1}{2} U^{(I)\,m'm''}_{mm''} \left[ \delta^m_{m'} - 2 n^{(I)(\sigma)m}_{m'} \right]. \end{aligned} \tag{7.20}$$

In the fully-averaged scalar interaction $U$ case this simplifies to

$$V^{(I)(\sigma)m'}_{m} = \frac{1}{2} U^{(I)} \left[ \delta^m_{m'} - 2 n^{(I)(\sigma)m}_{m'} \right], \tag{7.21}$$

and, in the intermediate singly orbital dependent case, it is given by

$$\begin{aligned} V^{(I)(\sigma)m'}_{m} &= \frac{1}{2} U^{(I)\,m'}_{m} \left[ \delta^m_{m'} - n_{m'}^{\ m''} \delta^m_{m''} - \delta^{m''}_{m'} n_{m''}^{\ m} \right]^{(I)(\sigma)} \\ &= \frac{1}{2} U^{(I)\,m'}_{m} \left[ \delta^m_{m'} - 2 n^{(I)(\sigma)m}_{m'} \right]. \end{aligned} \tag{7.22}$$

The equivalence of the potentials in the two-index and four-index cases demonstrates that the latter (the computation of which tensor involves the perturbation of individual potential matrix elements), while it may be of use in fully orbital-dependent functionals such as that shown in Eq. 2.7, is redundant for a simple DFT + $U$ term of the idempotency penalty-functional form.

### 7.1.6 Prospects for a Linear-Scaling Implementation

The independent-particle response is calculated during the process of computing the full response, and it does not excessively complicate the calculation. The usual strategy is that, beginning with the unperturbed ground-state density, the perturbing potential is added to the Hamiltonian and then the non-interacting energy is minimised with respect to the density with that perturbed Hamiltonian fixed.

The resulting change in the subspace occupancies is the independent particle contribution to be deducted from the Hubbard $U$. The Hamiltonian is then updated and brought to self-consistency with the density in order to obtain the full response, so that the non-interacting contribution is effectively computed at the first iteration of the self-consistency procedure.

This technique, of course, is more suited to the density-mixing approaches of conventional DFT methods that to the direct energy minimisation technique used in the linear-scaling code ONETEP. In the latter, it is not usually necessary to store the expansion of a fixed Hamiltonian on the basis (the *psinc* grid, for example) while carrying out energy minimisation, particularly during minimisation with respect to the support functions. This, however, is merely a technical obstacle and not one which fundamentally prohibits implementation of this method.

On the other hand, for a system consisting of a number of inequivalent correlated sites, it is necessary to separately carry out a number of linear scaling calculations in order to compute the response associated with perturbations at each site. These calculations should be, of course, very fast since the ground-state density is used as a starting point, but nonetheless the cost intrinsically scales with the number of correlated sites times the total number of atoms present.

## 7.2 The Constrained Random Phase Approximation

Moving next to our second example of an ab initio method for $U$, we seek to generalise the calculation of the Hubbard $U$ tensor using the Many-Body Perturbation Theory based approach known as the constrained Random Phase Approximation (cRPA) to the case of a nonorthogonal set of Hubbard projectors.

We formulate the methodology in the form of a nonorthogonal representation of single-particle states, such as NGWFs, and, where appropriate, we note issues pertinent to the potential for developing a linear-scaling implementation of the method. We closely follow the space–time *GW* methodology, prescribed in Ref. [16], for the construction of the screened Coulomb operator, and the description of cRPA itself which is provided in Ref. [4].

While we do not provide a detailed description of the derivation of the method, which is rather involved, we hope that the main steps in its construction will be made clear by our generalisation to nonorthogonal projectors. For further details on the method's development, we refer the reader to the seminal work of Ref. [17], in which many of the salient concepts were clarified, Ref. [4], the article in which the method was formally proposed, Ref. [9], in which the technique was generalised to include the exchange-correlation kernel for non-trivial interactions (for example to the cALDA—constrained Adiabatic Local Density Approximation) and successfully applied to a number of materials, and Ref. [18], in which cRPA was re-derived on rigorous foundations in terms of a dynamical Hubbard model.

The essence of the constrained Random Phase Approximation framework is to partition the polarisability propagation operator, usually computed from the Kohn–Sham system, into parts corresponding to transitions within the correlated subspace spanned by the Hubbard projectors on correlated site $I$ only, denoted $\hat{P}_{\mathcal{C}}^{(I)(\sigma)}$, and all others (including transitions between subspace and the bath with respect that particular subspace), denoted $\hat{P}_{\mathcal{B}}^{(I)(\sigma)} = \hat{P}^{(\sigma)} - \hat{P}_{\mathcal{C}}^{(I)(\sigma)}$. For the down-folded Hubbard model of the correlated subspace $\mathcal{C}^{(I)}$, the former contribution, $\hat{P}_{\mathcal{C}}^{(I)(\sigma)}$, should be excluded from the effective screened Coulomb operator.

The Random Phase Approximation (equivalent to a linearised time-dependent Hartree approximation and implying the neglect of the exchange-correlation kernel) is used to evaluate the dielectric function and hence the screened interaction. The fully screened interaction $\hat{W}$ may be expressed via the bath-screened $\hat{W}_{\mathcal{B}}^{(I)(\sigma)}$ (that which we need to compute the Hubbard $U$ tensor) and the bare Coulomb operator $\hat{v}$, via the Dyson equation-like expression

$$\hat{W}^{(I)(\sigma)} = \left(\hat{1} - \hat{W}_{\mathcal{B}}^{(I)(\sigma)} \hat{P}_{\mathcal{C}}^{(I)(\sigma)}\right)^{-1} \hat{W}_{\mathcal{B}}^{(I)(\sigma)}, \quad \text{where}$$
$$\hat{W}_{\mathcal{B}}^{(I)(\sigma)} = \left(\hat{1} - \hat{v} \hat{P}_{\mathcal{B}}^{(I)(\sigma)}\right)^{-1} \hat{v}, \tag{7.23}$$

which makes explicit the interpretation of the interaction $\hat{W}_{\mathcal{B}}^{(I)(\sigma)}$ as a partially screened intermediate between $\hat{v}$ and $\hat{W}^{(I)(\sigma)}$.

In the process of computing the Hubbard $U$, which is energy-dependent in this theory, one obtains the independent-particle spectral function as a useful by-product, along with much of the machinery needed for $GW$ [19–23] many-body corrections to the Kohn–Sham spectrum. We show how the frequency dependence of the Hubbard $U$ may be used to make a simple first-order correction to the DFT + $U$ spectral function for dynamical correlation effects in the spirit of the perturbative $GW$ method.

The computed Hubbard $U$, in its static limit, may be used to re-build the DFT + $U$ correction, as demonstrated in Ref. [5]. Furthermore, it may, in principle, be coupled to our projector self-consistent DFT + $U$ method, detailed in Chap. 3 and Ref. [10], in order to attain self-consistency over the charge density, Hubbard projectors and Hubbard interaction parameters.

### 7.2.1 *The Independent-Particle Green's Function and Irreducible Polarisability Operator*

We begin with the Kohn–Sham (non-interacting) Green's function, which describes the probability amplitude for single-particle propagation at a given energy. It may be expressed in real-space and frequency, for a given spin-channel (where the infinitesimal $\delta$ provides a Lorentzian broadening of the poles in order to ensure invertibility) by

$$G^{(\sigma)}\left(\mathbf{r}, \mathbf{r}'; \omega\right) = \langle \mathbf{r} | \left( (\omega + \imath\delta)\, \hat{1} + \hat{H}^{(\sigma)} \right)^{-1} | \mathbf{r}' \rangle \tag{7.24}$$

In the NGWF representation, the Green's function is most naturally expressed as a doubly contravariant tensor, computed over a frequency range encapsulating twice the Kohn–Sham bandwidth, by inverting the covariant representation of the inverse Green's function given by

$$\bar{G}^{(\sigma)}_{\alpha\beta}(\omega) = (\omega + \iota\delta) S_{\alpha\beta} + H^{(\sigma)}_{\alpha\beta}; \quad G^{(\sigma)\alpha\beta} = \left(\bar{G}^{(\sigma)-1}\right)^{\alpha\beta}. \tag{7.25}$$

The inversion of the Green's function with linear-scaling cost is something of a challenge for energies close to its poles, and it may prove that simple recursive product methods such as Hotelling's algorithm [24], for example, may require an excessive value for the broadening parameter $\delta$ for accurate results. Further investigation into methods for inversion of sparse, but close to singular, matrices is required for routine linear-scaling applications of Green's function methods.

Within the Random Phase Approximation (which includes dynamical effects but neglects all but the Hartree interaction between particles and holes), the irreducible polarisability propagator is given by the convolution of the Green's function with itself, that is in matrix language and ignoring spin-flip excitations,

$$P^{(\sigma)\alpha\beta}(\omega) = -\frac{\iota}{2\pi}\int_{-\infty}^{\infty} d\omega' G^{(\sigma)\alpha\gamma}\left(\omega'\right) S_{\gamma\epsilon} G^{(\sigma)\epsilon\beta}\left(\omega' - \omega\right). \tag{7.26}$$

In order to make the connection between the cRPA and linear density-response theories clearer, we note that the retarded (computed using DFT or TDDFT) and time-ordered (computed using many-body perturbation theory) response functions become equivalent, so that $\hat{P}(\omega \to 0) \equiv \hat{\chi}(\omega \to 0)$, in the static limit appropriate for building the DFT + $U$ parameters.

From a computational point of view, it is simpler to compute the inverse Fourier transform of the Green's function to real time,

$$G^{(\sigma)\alpha\beta}(t) = \frac{1}{2\pi}\int_{-\infty}^{\infty} d\omega e^{+\iota\omega t} G^{(\sigma)\alpha\gamma}(\omega), \tag{7.27}$$

after which the polarisability may be computed using the transform of a simple square of the time-domain Green's function, i.e.,

$$P^{(\sigma)\alpha\beta}(\omega) = -\iota\int_{-\infty}^{\infty} dt e^{-\iota\omega t} G^{(\sigma)\alpha\gamma}(t) S_{\gamma\epsilon} G^{(\sigma)\epsilon\beta}(t), \tag{7.28}$$

which computationally simplifies to a sum of products in the static limit.

We note that using real time and frequency domains may not be so computationally efficient or robust as using the imaginary time domain, since the quantities involved are much smoother in the latter case [16]. However, we present the formalism on the real axis for clarity. A number of methods have been developed for efficient evaluation of the polarisability operator in the context of ab initio simulation, in particular we refer the reader to Ref. [25] for a very elegant approach.

### 7.2.2 Spectral Functions

Let us briefly mention some quantities which may already be computed once we have come this far. The density of Kohn–Sham states is equivalent to the spectral function of the Green's function, that is

$$\mathrm{DoS}^{(\sigma)}(\omega) = \frac{1}{\pi} sgn(\omega) \Im \left[ G^{(\sigma)\alpha\beta}(\omega) \right] S_{\beta\alpha}. \tag{7.29}$$

Oscillator strengths for independent particle transitions (equivalent to Fermi's Golden Rule) are provided by the spectral function of the polarisability operator, that is

$$A^{(\sigma)}(\omega) = \frac{1}{\pi} sgn(\omega) \Im \left[ P^{(\sigma)\alpha\beta}(\omega) \right] S_{\beta\alpha}. \tag{7.30}$$

Somewhat more interesting, however, is the weighting of such oscillator strengths by appropriate moments in order to obtain frequency-dependent polarisabilities, the simplest being the electric dipole–dipole polarisability

$$\alpha_{ij}^{(\sigma)}(\omega) = \frac{1}{\pi} sgn(\omega) \langle \phi_\alpha | \mathbf{r}_i | \phi_\beta \rangle P^{(\sigma)\beta\gamma}(\omega) \langle \phi_\gamma | \mathbf{r}_j | \phi_\delta \rangle S^{\delta\alpha}, \tag{7.31}$$

from which the corresponding cross-section for single-particle excitations (photoemission or inverse photoemission) may be computed using the expression

$$\sigma_{ij}(\omega) = \frac{4\pi}{c} \omega \Im \alpha_{ij}^{(\sigma)}(\omega). \tag{7.32}$$

An advantage of computing the polarisability in a frame of localised Wannier functions, of course, is that we may conveniently decompose it into its contributions from transitions between functions with different symmetries or host atoms simply by partially evaluating the sums in Eq. 7.31.

The generalisation to rotatory (chiroptical) cross-sections and the inclusion of dynamical excitation effects are very interesting topics, involving the replacement of the position operators by angular momentum operators in the above expressions and the use of TDDFT, respectively; we refer the reader to fascinating articles on these topics at Refs. [26, 27] and Refs. [14, 28].

### *7.2.3 The Low-Energy Hubbard Model of cRPA*

In order to build the parameters of a Hubbard model only of the correlated subspace, we must screen all of the Coulomb interactions involving the bath with respect to that subspace (the theory allows for one inequivalent Hubbard site to be parameterised at a time). We first compute the polarisability operator for low-energy single-particle transitions within the subspace, $\hat{P}_C^{(I)(\sigma)}$, by taking the real-time product of subspace-projected Green's functions given by

$$G_{C\ \ m}^{(I)(\sigma)m'}(t) = \langle \varphi_m^{(I)} | \hat{G}^{(\sigma)}(t) | \varphi^{(I)m'} \rangle, \tag{7.33}$$

which, in the NGWF representation of Chap. 2, are expressed as

$$G_{\mathcal{C}\ \ m}^{(I)(\sigma)m'}(t) = W_{m\alpha}^{(I)} G^{(\sigma)\alpha\beta}(t)\, V_{\beta m''}^{(I)} O^{(I)m''m'}. \tag{7.34}$$

Here we employ the transformation matrix $V^{(I)}{}_{\alpha m} = \langle \phi_\alpha | \varphi^{(I)}{}_m \rangle$, its transpose $W^{(I)}{}_{m\alpha} = \langle \varphi^{(I)}{}_m | \phi_\alpha \rangle$ and the contravariant (with respect to the correlated subspace only—see Chap. 4 and Ref. [29]) metric tensor $O^{(I)mm'} = \langle \varphi^{(I)m} | \varphi^{(I)m'} \rangle$.

The product of Green's functions may be carried out in the localised frame of Hubbard projectors, via the matrix product (where the time domain argument is understood and the function indices are suppressed for brevity)

$$\iota P_{\mathcal{C}}^{(I)(\sigma)} = S V^{(I)} O^{(I)} G_{\mathcal{C}}^{(I)(\sigma)} G_{\mathcal{C}}^{(I)(\sigma)} W^{(I)} S. \tag{7.35}$$

The required contravariant irreducible bath polarisability is then given by the difference of the full and subspace polarisabilities,

$$\iota P_{\mathcal{B}}^{(I)(\sigma)} = G^{(\sigma)} S G^{(\sigma)} - S V^{(I)} O^{(I)} G_{\mathcal{C}}^{(\sigma)(I)} G_{\mathcal{C}}^{(\sigma)(I)} W^{(I)} S. \tag{7.36}$$

### 7.2.4 Dielectric Function, Screened Coulomb Interaction and Hubbard *U* Tensor

In order to compute the Hubbard $U$ tensor using cRPA, we must first locate the screened Coulomb interaction of the bath with respect to the correlated subspace at hand. The dynamically screened Coulomb interaction may be computed, in terms of the bare Hartree potential,

$$v\left(\mathbf{r}, \mathbf{r}'\right) = \frac{1}{|\mathbf{r} - \mathbf{r}'|}. \tag{7.37}$$

and the inverse of the Lindhard dielectric function, $\epsilon(\omega)$. The latter may be expressed in real or reciprocal space, respectively, as

$$W\left(\mathbf{r}, \mathbf{r}'; \omega\right) = \int d^3\mathbf{r}''\, v\left(\mathbf{r}, \mathbf{r}''\right) \epsilon^{-1}\left(\mathbf{r}'', \mathbf{r}'; \omega\right), \tag{7.38}$$

or, considering Gamma-point sampling only for large systems,

$$W\left(\mathbf{G}, \mathbf{G}'; \omega\right) = \frac{4\pi}{|\mathbf{G}|\,|\mathbf{G}'|} \epsilon^{-1}\left(\mathbf{G}, \mathbf{G}'; \omega\right). \tag{7.39}$$

In order to calculate the dielectric matrix, we invoke the RPA expression for the dielectric function

$$\epsilon\left(\mathbf{r}, \mathbf{r}'; \omega\right) = \delta\left(\mathbf{r} - \mathbf{r}'\right) - \int d^3\mathbf{r}''\, v\left(\mathbf{r}, \mathbf{r}''\right) P\left(\mathbf{r}'', \mathbf{r}'; \omega\right), \tag{7.40}$$

which, of course, also has convenient expression in reciprocal space as

$$\epsilon^{(\sigma)}\left(\mathbf{G},\mathbf{G}';\omega\right)=\delta_{\mathbf{G}\mathbf{G}'}-\frac{4\pi}{|\mathbf{G}|\,|\mathbf{G}'|}P\left(\mathbf{G},\mathbf{G}';\omega\right). \tag{7.41}$$

One possible way to proceed is by directly computing $\epsilon_{\mathcal{B}}^{(I)(\sigma)}$ on the FFTBOX grid, at each frequency required, using the formula

$$\epsilon_{\mathcal{B}}^{(I)(\sigma)}\left(\mathbf{G},\mathbf{G}'\right)=\delta_{\mathbf{G}\mathbf{G}'}-\frac{4\pi}{|\mathbf{G}|\,|\mathbf{G}'|}\phi_{\alpha}\left(\mathbf{G}\right)P_{\mathcal{B}}^{(I)(\sigma)\alpha\beta}\phi_{\beta}\left(\mathbf{G}'\right). \tag{7.42}$$

It may prove more efficient for large systems, however, to use the NGWF expansion of the Hartree potential, that is

$$v_{\alpha\beta}=\int\int d^{3}\mathbf{r}d^{3}\mathbf{r}'\phi_{\alpha}\left(\mathbf{r}\right)\frac{1}{|\mathbf{r}-\mathbf{r}'|}\phi_{\beta}\left(\mathbf{r}'\right), \tag{7.43}$$

in order to directly evaluate the matrix elements of the dielectric function via

$$\epsilon_{\mathcal{B}}^{(I)(\sigma)\beta}{}_{\alpha}\left(\omega\right)=1_{\alpha}{}^{\beta}-v_{\alpha\gamma}P_{\mathcal{B}}^{(I)(\sigma)\gamma\beta}\left(\omega\right), \tag{7.44}$$

whence the matrix elements of the screened bath Coulomb interaction are expressed as

$$W_{\mathcal{B}\alpha\beta}^{(\sigma)}\left(\omega\right)=\left(1_{\alpha}{}^{\gamma}-v_{\alpha\delta}P_{\mathcal{B}}^{(I)(\sigma)\delta\gamma}\left(\omega\right)\right)^{-1}v_{\gamma\beta}. \tag{7.45}$$

The evaluation of the polarisability propagator, dielectric function and screened Coulomb interaction directly on the *psinc* grid brings the advantage of NGWF independence—no loss of accuracy is introduced by the introduction of a spatially localised representation. On the other hand, if matrix sparsity is adequately accounted for, the latter should not introduce any detrimental approximations. Let us denote the sparsity of the density kernel and the NGWF overlap matrix by **K** and **S**, respectively. The Green's function, by construction, has a sparsity pattern similar to the density kernel, and so the polarisability operator has a sparsity pattern of **KSK**. The subtraction of the subspace polarisability does not alter this.

The bare Coulomb interaction may be reasonably assumed to be adequately described by the sparsity of the Hamiltonian, **H**, so the dielectric function is fully described by a matrix of sparsity **HKSK**. Of course, the dielectric function is, in principle, further delocalised upon inversion but, supposing that it may not be necessary to include a larger number of matrix elements in practice, the screened Coulomb interaction then is adequately described by a matrix of sparsity pattern **HKSKH**. Only a small block of this matrix, of sparsity **H**, is actually needed for the Hubbard $U$ tensor, so further economies may, of course, be introduced if $W_{\mathcal{B}\alpha\beta}^{(\sigma)}(\omega)$ is not further required.

The irreducible polarisability propagator, and thus the dielectric function and screened Coulomb interaction, require expansion on a *psinc* grid twice as fine as that on which the Green's function is represented in order to capture all of the necessary

Fourier components. Furthermore, the NGWF representation, as it is currently optimised, is adequate only for conduction states lying close to the Fermi level. Additional NGWFs should be optimised for a substantial number of conduction-band states, or simply augmented with an appropriate atomic basis, in order to satisfactorily represent the dielectric function in practice.

Supposing that these technical obstacles may be overcome, then a dynamical and fully orbital-resolved Hubbard $U$ tensor is provided by a localised four-centre integral of the bath-screened interaction $W^{(\sigma)}_{\mathcal{B}\alpha\beta}(\omega)$ over the Hubbard projectors, that is (for a doubly covariant and doubly contravariant example)

$$U^{(I)(\sigma)m'm'''}_{\quad mm''}(\omega) = \langle \varphi^{(I)}_{m} \varphi^{(I)m'} | \hat{W}^{(I)(\sigma)}_{\mathcal{B}}(\omega) | \varphi^{(I)}_{m''} \varphi^{(I)m'''} \rangle, \tag{7.46}$$

where only the diagonal Coulomb repulsion elements, whereat $m = m''$; $m' = m'''$, and the static limit $\omega \rightarrow 0$ (where $U$ is real-valued) are typically of interest for re-building the DFT + $U$ correction.

### 7.2.5 Making Use of a Frequency-Dependent U

We cannot exploit the energy dependence of the Hubbard $U$ in Hamiltonian formalisms such as DFT + $U$. Notwithstanding, one may put this frequency dependence to use in order to compute dynamical corrections to the DFT + $U$ Green's function in a perturbative manner. Such corrections could conceivably be made, say, at self-consistency with the static ($\omega \rightarrow 0$) limit of the $U$ as demonstrated in Ref. [5].

One may, for example, re-build the Green's function (and hence the single-particle spectra) by correcting the ground-state Hamiltonian corrected for dynamical effects—so that it becomes a model self-energy operator in a similar approach to that which is often used to approximately correct the eigenvalue differences to first-order using the $GW$ self-energy operator, while retaining the uncorrected Kohn–Sham eigenstates.

The correction to be used is simply the DFT + $U$ potential operator scaled by the frequency-dependent $U$ (we consider a scalar parameter in this example, for notational simplicity), that is

$$\Delta^{(\sigma)DFT+U}_{\alpha\beta}(\omega) = \sum_{I} \left( U^{(I)(\sigma)}(\omega) - U^{(I)(\sigma)}(\omega = 0) \right) C^{(I)(\sigma)}_{\alpha\beta};$$
$$C^{(\sigma)(I)}_{\alpha\beta} = \frac{1}{2} V^{(I)}_{\alpha m} \left( \bar{O} - 2 O W K^{(\sigma)} V O \right)^{(I)mm'} W^{(I)}_{m'\beta}. \tag{7.47}$$

The inverse Green's function, and hence the spectral function, is then modified, non self-consistently, for the frequency-dependence of the Hubbard $U$, according to

$$\left[ \bar{G}^{(\sigma)} \right]_{\alpha\beta}(\omega) = (\omega + \iota\delta) S_{\alpha\beta} + H^{(\sigma)DFT+U(\omega=0)}_{\alpha\beta} + \Delta^{(\sigma)DFT+U}_{\alpha\beta}. \tag{7.48}$$

As a final remark on this topic, we note that most of the quantities needed to compute the $GW$ corrections to the quasiparticle energies have already been described above; $GW$ and cRPA are constructed in very similar ways. In the simplest approximations to the $GW$ method, known as $G^0 W^0$, the self-energy is expanded to first order in the screened Coulomb interaction, while retaining the Kohn–Sham Green's function and $W$, thereby using the self-energy as a correction to the exchange-correlation potential.

The self-energy operator is then given, as a point-wise spatial multiplication in the time domain, by

$$\hat{\Sigma}^{(\sigma)}\left(\mathbf{r}, \mathbf{r}'; t\right) = \iota \hat{G}^{(\sigma)}\left(\mathbf{r}, \mathbf{r}'; t\right) \hat{W}^{(\sigma)}\left(\mathbf{r}, \mathbf{r}'; t + \delta\right), \tag{7.49}$$

where a Green's function, first corrected for the dynamical DFT + $U$ correlations which we have proposed, may be used in place of the non-interacting Green's function, if so desired. The covariant NGWF representation of the self-energy in the energy domain is given by

$$\Sigma^{(\sigma)}_{\alpha\beta}\left(\omega\right) = \int_{-\infty}^{\infty} dt e^{-\iota\omega t} \int d^3\mathbf{r} d^3\mathbf{r}' \phi_\alpha\left(\mathbf{r}\right) \hat{\Sigma}^{(\sigma)}\left(\mathbf{r}, \mathbf{r}'; t\right) \phi_\beta\left(\mathbf{r}'\right). \tag{7.50}$$

## 7.3 Interaction Tensor Update with Hubbard Projectors

In this section, we derive the expressions needed to compute the update to the Coulomb interaction tensor which is concomitant, to first order, with changes in the set of Hubbard projectors with respect to which it is defined. This is of importance to a combined Hubbard projector and interaction tensor self-consistent DFT + $U$ method since, for example, if the projectors are close to consistency then it may prove to be preferable to estimate changes to the interaction tensor using the expressions which we go on to provide instead of an expensive re-computation of the Hubbard $U$ tensor at each projector update iteration.

On the other hand, if the Hubbard projectors are themselves computed by maximising the interaction tensor or by minimising its anisotropy, by some appropriate measure, then it is necessary to employ the expressions derived in this section if the projectors are nonorthogonal.

We make use of the geometric principles introduced in Chap. 5 in order to ensure that the tensorial integrity of all quantities is preserved and we accommodate interaction tensors both of rank-2 and those fully projector-resolved interactions of rank-4.

### 7.3.1 Geometry of the Hubbard Support Manifolds

As we have discussed in Chap. 4 and Ref. [29], distinct (though potentially overlapping) correlated subspaces must be imbued with individual metric tensors, that is an individual $O^{(I)}_{\bullet\bullet}$ for each subspace in the covariant representation, if the tensorial invariance of the total energy is to be ensured. As a result, there is a distinct nontrivial support manifold, call it a *Hubbard support manifold*, and a corresponding Levi-Civita connection (representation-independent definition of differentiation) associated with each correlated subspace.

As a consequence of the arguments given in Chap. 5, many of the results for the geometry of the principal support manifold also apply to the Hubbard support manifolds. For example, the correlated subspaces spanned by real-valued Hubbard projectors exhibit torsion free, hyperbolic geometries.

Applying the same arguments used in our treatment of the support manifold for the Kohn–Sham states, to the case of the support manifolds for the strongly correlated states, we invoke the approximation that the correlated subspace spanned by each set of real-valued Hubbard projectors does not change to first order and so the derivatives of the identity operator on those subspaces vanishes.

Thus, we separately compute the partial derivative of the completeness relationship,

$$\varphi^{(I)}_{m} O^{(I)mm'} \varphi^{(I)}_{m'} = \hat{1}^{(I)} = \varphi^{(I)m} O^{(I)}_{mm'} \varphi^{(I)m'}, \tag{7.51}$$

with respect to the Hubbard projectors for each correlated subspace, contravariant and covariant, providing, respectively, that

$$O^{(I)mm',m''} = -\varphi^{(I)m} O^{(I)m'm''} - \varphi^{(I)m'} O^{(I)mm''} \quad \text{and} \tag{7.52}$$

$$O^{(I)}_{mm',m''} = -\varphi^{(I)}_{m} O^{(I)}_{m'm''} - \varphi_{m'} O^{(I)}_{mm''}. \tag{7.53}$$

From here, we may construct the Levi-Civita connection on the Hubbard support manifold (the representation-independent definition of the absolute derivative which leaves the metric $O_{\bullet\bullet}$ invariant to first order) in an identical manner to our argument for the principal support manifold.

The Christoffel symbol of the first and second kinds are given, for each subspace $I$, by

$$\Gamma^{(I)}_{mm'm''} = -\varphi^{(I)}_{m} O^{(I)}_{m'm''} \quad \text{and} \tag{7.54}$$

$$\Gamma^{(I)m}_{\quad m'm''} = O^{(I)m'm'''} \Gamma^{(I)}_{m'''m'm''} = -\varphi^{(I)m} O^{(I)}_{m'm''}. \tag{7.55}$$

Henceforth, for notational clarity, we suppress the subspace index $I$ and employ Greek indices, with the understanding that all implicit summation runs over the Hubbard projector indices relevant to the correlated subspace in question only.

### 7.3.2 First Order Changes to the Hubbard U Tensor

Using the Christoffel symbols, we may, for example, calculate the covariant absolute derivative of the fully contravariant rank-2 Hubbard interaction tensor which is given by

$$U^{\alpha\beta}_{;\gamma} = U^{\alpha\beta}_{,\gamma} + \Gamma^{\alpha}_{\gamma\delta} U^{\delta\beta} + \Gamma^{\beta}_{\gamma\delta} U^{\alpha\delta}. \tag{7.56}$$

That of rank-4 is given by the slightly more complicated expression

$$\begin{aligned} U^{\alpha\beta\gamma\delta}_{;\epsilon} &= U^{\alpha\beta\gamma\delta}_{,\epsilon} + \Gamma^{\alpha}_{\epsilon\zeta} U^{\zeta\beta\gamma\delta} + \Gamma^{\beta}_{\epsilon\zeta} U^{\alpha\zeta\gamma\delta} \\ &+ \Gamma^{\gamma}_{\epsilon\zeta} U^{\alpha\beta\zeta\delta} + \Gamma^{\delta}_{\epsilon\zeta} U^{\alpha\beta\gamma\zeta}. \end{aligned} \tag{7.57}$$

In order to find the first-order change in the interaction tensor due to a change in the Hubbard projectors we recall Ricci's Lemma, which states that covariant differentiation commutes with the metric tensor. In the case of a rank-2 interaction tensor, say of form $U_{\bullet}{}^{\bullet}$, the change to the tensor is

$$\begin{aligned} \delta U_{\alpha}{}^{\beta} &= \int d\mathbf{r}\; U_{\alpha}{}^{\beta;\gamma}(\mathbf{r})\, \delta\varphi_{\gamma}(\mathbf{r}) \\ &= \int d\mathbf{r}\; O_{\alpha\delta} U^{\delta\beta}_{;\epsilon}(\mathbf{r})\, O^{\epsilon\gamma} \delta\varphi_{\gamma}(\mathbf{r}) \\ &= \int d\mathbf{r}\; O_{\alpha\delta} \left( U^{\delta\beta}_{,\epsilon} + \Gamma^{\delta}_{\epsilon\zeta} U^{\zeta\beta} + \Gamma^{\beta}_{\epsilon\zeta} U^{\delta\zeta} \right)(\mathbf{r})\, O^{\epsilon\gamma} \delta\varphi_{\gamma}(\mathbf{r}) \\ &= \int d\mathbf{r}\; O_{\alpha\delta} \left( U^{\delta\beta}_{,\epsilon} - \varphi^{\delta} O_{\epsilon\zeta} U^{\zeta\beta} - \varphi^{\beta} O_{\epsilon\zeta} U^{\delta\zeta} \right)(\mathbf{r})\, O^{\epsilon\gamma} \delta\varphi_{\gamma}(\mathbf{r}) \\ &= \int d\mathbf{r}\; \left( O_{\alpha\delta} U^{\delta\beta}_{,\epsilon}(\mathbf{r})\, O^{\epsilon\gamma} - \varphi_{\alpha}(\mathbf{r})\, U^{\gamma\beta} - \varphi^{\beta}(\mathbf{r})\, U_{\alpha}{}^{\gamma} \right) \delta\varphi_{\gamma}(\mathbf{r}). \end{aligned} \tag{7.58}$$

Here we have chosen to cast the tensor in its fully contravariant form before expressing the covariant derivative in its component parts so that the partial derivative may be most conveniently evaluated.

Similarly, for the case of a rank-4 interaction tensor, say, for example, of the form $U_{\bullet}{}^{\bullet\bullet}{}_{\bullet}$, the partial derivative with respect to the contravariant vectors may be most easily evaluated by first casting the tensor into its fully contravariant form, so we compute the first order change in the tensor via the expression

$$
\begin{aligned}
\delta U_{\alpha}{}^{\beta\gamma}{}_{\delta} &= \int d\mathbf{r}\; U_{\alpha}{}^{\beta\gamma;\epsilon}{}_{\delta}\left(\mathbf{r}\right)\delta\varphi_{\epsilon}\left(\mathbf{r}\right) \\
&= \int d\mathbf{r}\; O_{\alpha\zeta}O_{\delta\eta}U^{\zeta\beta\gamma\eta}_{;\theta}\left(\mathbf{r}\right)O^{\theta\epsilon}\delta\varphi_{\epsilon}\left(\mathbf{r}\right) \\
&= \int d\mathbf{r}\; O_{\alpha\zeta}O_{\delta\eta}\begin{pmatrix} U^{\zeta\beta\gamma\eta}_{,\theta} \\ +\Gamma^{\zeta}_{\theta\iota}U^{\iota\beta\gamma\eta}+\Gamma^{\beta}_{\theta\iota}U^{\zeta\iota\gamma\eta} \\ +\Gamma^{\gamma}_{\theta\iota}U^{\zeta\beta\iota\eta}+\Gamma^{\eta}_{\theta\iota}U^{\zeta\beta\gamma\iota}\end{pmatrix}(\mathbf{r})O^{\theta\epsilon}\delta\varphi_{\epsilon}(\mathbf{r}) \\
&= \int d\mathbf{r}\begin{pmatrix} O_{\alpha\zeta}O_{\delta\eta}U^{\zeta\beta\gamma\eta}_{,\theta}O^{\theta\epsilon} \\ -\varphi_{\alpha}U^{\epsilon\beta\gamma}{}_{\delta}-\varphi^{\beta}U_{\alpha}{}^{\epsilon\gamma}{}_{\delta} \\ -\varphi^{\gamma}U_{\alpha}{}^{\beta\epsilon}{}_{\delta}-\varphi_{\delta}U_{\alpha}{}^{\beta\gamma\epsilon}\end{pmatrix}(\mathbf{r})\delta\varphi_{\epsilon}(\mathbf{r}).
\end{aligned}
\tag{7.59}
$$

In principle, these expressions are not difficult to compute if the partial derivative of the interaction tensor is available. Let us first consider the case where the (4-index) interaction tensor is given by a two-centre integral of a screened Coulomb interaction operator $\hat{W}$, which may be frequency-dependent but not explicitly dependent on the Hubbard projectors themselves. Such a tensor is generally defined in Dirac notation by

$$
\begin{aligned}
U^{\zeta\beta\gamma\eta} &= \langle\varphi^{\zeta}\varphi^{\beta}|\hat{W}|\varphi^{\gamma}\varphi^{\eta}\rangle \\
&= \int d\mathbf{r}\int d\mathbf{r}'\;\varphi^{*\alpha}\left(\mathbf{r}\right)\varphi^{*\beta}\left(\mathbf{r}'\right)\hat{W}\left(\mathbf{r},\mathbf{r}'\right)\varphi^{\gamma}\left(\mathbf{r}\right)\varphi^{\delta}\left(\mathbf{r}'\right).
\end{aligned}
\tag{7.60}
$$

Considering real-valued functions only, as we are obliged to while remaining in the realm of simple metric-compatible support manifolds, the partial derivative with respect to support functions is provided by the sum of integrals

$$
\begin{aligned}
U^{\zeta\beta\gamma\eta}_{,\theta} &= \delta^{\zeta}_{\theta}\langle\varphi^{\beta}|\hat{W}|\varphi^{\gamma}\varphi^{\eta}\rangle+\delta^{\beta}_{\theta}\langle\varphi^{\zeta}|\hat{W}|\varphi^{\gamma}\varphi^{\eta}\rangle \\
&\quad+\delta^{\gamma}_{\theta}\langle\varphi^{\zeta}\varphi^{\beta}|\hat{W}|\varphi^{\eta}\rangle+\delta^{\eta}_{\theta}\langle\varphi^{\zeta}\varphi^{\beta}|\hat{W}|\varphi^{\gamma}\rangle \\
&= \left(\delta^{\zeta}_{\theta}U^{\iota\beta\gamma\eta}+\delta^{\beta}_{\theta}U^{\zeta\iota\gamma\eta}+\delta^{\gamma}_{\theta}U^{\zeta\beta\iota\eta}+\delta^{\eta}_{\theta}U^{\zeta\beta\gamma\iota}\right)\varphi_{\iota}\left(\mathbf{r}\right),
\end{aligned}
\tag{7.61}
$$

and we conclude that the required term in the first order update to $U$ is given by

$$
O_{\alpha\zeta}O_{\delta\eta}U^{\zeta\beta\gamma\eta}_{,\theta}O^{\theta\epsilon}=\begin{pmatrix}\delta^{\epsilon}_{\alpha}U^{\iota\beta\gamma}{}_{\delta}+O^{\beta\epsilon}U_{\alpha}{}^{\iota\gamma}{}_{\delta} \\ +O^{\gamma\epsilon}U_{\alpha}{}^{\beta\iota}{}_{\delta}+\delta^{\epsilon}_{\delta}U_{\alpha}{}^{\beta\gamma\iota}\end{pmatrix}\varphi_{\iota}(\mathbf{r}).
\tag{7.62}
$$

Using this, and noting an elegant antisymmetry between terms due to partial derivatives and Christoffel symbols, we can most succinctly express the first-order change in the interaction tensor due to Hubbard projector update as

$$
\begin{aligned}
\delta U_{\alpha}{}^{\beta\gamma}{}_{\delta} &= \left(\langle\delta\varphi_{\alpha}|\varphi_{\epsilon}\rangle-\langle\varphi_{\alpha}|\delta\varphi_{\epsilon}\rangle\right)U^{\epsilon\beta\gamma}{}_{\delta} \\
&\quad+\left(\langle\delta\varphi^{\beta}|\varphi_{\epsilon}\rangle-\langle\varphi^{\beta}|\delta\varphi_{\epsilon}\rangle\right)U_{\alpha}{}^{\epsilon\gamma}{}_{\delta} \\
&\qquad+\left(\langle\delta\varphi^{\gamma}|\varphi_{\epsilon}\rangle-\langle\varphi^{\gamma}|\delta\varphi_{\epsilon}\rangle\right)U_{\alpha}{}^{\beta\epsilon}{}_{\delta} \\
&\qquad\quad+\left(\langle\delta\varphi_{\delta}|\varphi_{\epsilon}\rangle-\langle\varphi_{\delta}|\delta\varphi_{\epsilon}\rangle\right)U_{\alpha}{}^{\beta\gamma\epsilon}.
\end{aligned}
\tag{7.63}
$$

The corresponding expression in the rank-2 case is provided by the somewhat simpler equation

$$\delta U_{\alpha}{}^{\gamma} = \left(\langle \delta\varphi_{\alpha} | \varphi_{\epsilon}\rangle - \langle \varphi_{\alpha} | \delta\varphi_{\epsilon}\rangle\right) U^{\epsilon\gamma} + U_{\alpha}{}^{\epsilon} \left(\langle \varphi_{\epsilon} | \delta\varphi^{\gamma}\rangle - \langle \delta\varphi_{\epsilon} | \varphi^{\gamma}\rangle\right). \quad (7.64)$$

We may assure ourselves that contraction over two indices in the rank-4 expression results in cancellation of terms, so that $\delta U_{\alpha}{}^{\beta\gamma}{}_{\beta} = \delta U_{\alpha}{}^{\gamma}$, as expected. Furthermore, the latter expression is readily seen to be trace-free, as we would expect since the preservation of the subspace spanned by the Hubbard projectors ensures that the trace of the interaction tensor is invariant. As a result, the change in the scalar invariant interaction

$$\delta U = \delta U_{\alpha}{}^{\alpha} = \delta U_{\alpha}{}^{\beta\alpha}{}_{\beta} = 0 \quad (6.65)$$

to first order in changes of the Hubbard projectors when the interaction operator does not change with the projectors.

### *7.3.3 Invariance of the Interaction Anisotropy*

Some alternative scalar invariants, to the aforementioned trace, include

$$\begin{aligned} \tilde{U} &= U_{\alpha}{}^{\alpha\beta}{}_{\beta}, \\ J &= U_{\alpha}{}^{\beta\gamma}{}_{\delta} O_{\gamma\beta} O^{\delta\alpha} \quad \text{and} \\ (\Delta U)^2 &= \frac{3}{2} U_{\alpha}{}^{\beta} U_{\beta}{}^{\alpha} - \frac{1}{2}\left(U_{\alpha}{}^{\alpha}\right)^2 = \frac{3}{2} U_{\alpha}{}^{\gamma\beta}{}_{\gamma} U_{\beta}{}^{\gamma\alpha}{}_{\gamma} - \frac{1}{2}\left(U_{\alpha}{}^{\gamma\alpha}{}_{\gamma}\right)^2, \end{aligned} \quad (7.66)$$

which represent a measure of the non-local part of the interaction, the averaged same-spin exchange interaction and the square of the anisotropy of the rank-2 interaction tensor, respectively. While the first two scalars may be of general interest, we are more concerned with local Coulomb repulsion interactions and so we do not explore the projector-dependence of these objects further.

We briefly comment only on the latter tensorial invariant, the anisotropy of the two-index Coulomb interaction. An interesting question is whether there is any physical significance to the set of Hubbard projectors, spanning a given correlated subspace, which provide a maximally isotropic (or minimally anisotropic) interaction tensor. In that sense, such projectors would represent the optimally spherical representation of the projection of the interaction operator onto that subspace.

However, the integral of the covariant derivative of the squared scalar anisotropy with a test change vector vanishes, because

$$
\begin{aligned}
\delta (\Delta U)^2 &= \int d\mathbf{r} \; (\Delta U)^2_{;\epsilon} (\mathbf{r}) \, \delta\varphi^{\epsilon} (\mathbf{r}) \\
&= \int d\mathbf{r} \left( \frac{3}{2} U_{\alpha}{}^{\beta}{}_{;\epsilon} U_{\beta}{}^{\alpha} + \frac{3}{2} U_{\alpha}{}^{\beta} U_{\beta}{}^{\alpha}{}_{;\epsilon} - U_{\alpha}{}^{\alpha} U_{\beta}{}^{\beta}{}_{;\epsilon} \right) (\mathbf{r}) \, \delta\varphi^{\epsilon} (\mathbf{r}) \\
&= \int d\mathbf{r} \left( 3 U_{\alpha}{}^{\beta}{}_{;\epsilon} U_{\beta}{}^{\alpha} \right) (\mathbf{r}) \, \delta\varphi^{\epsilon} (\mathbf{r}) \\
&\quad \left( since \int d\mathbf{r} \; U_{\beta}{}^{\beta}{}_{;\epsilon} (\mathbf{r}) \, \delta\varphi^{\epsilon} (\mathbf{r}) = 0 \right) \\
&= 3 O_{\alpha\delta} U_{\beta}{}^{\alpha} \int d\mathbf{r} \; U^{\delta\beta}_{;\epsilon} (\mathbf{r}) \, \delta\varphi^{\epsilon} (\mathbf{r}) \\
&= 3 O_{\alpha\delta} U_{\beta}{}^{\alpha} \int d\mathbf{r} \begin{pmatrix} U^{\delta\beta}_{,\epsilon} (\mathbf{r}) - \varphi^{\delta} (\mathbf{r}) \, O_{\epsilon\zeta} U^{\zeta\beta} \\ -\varphi^{\beta} (\mathbf{r}) \, O_{\epsilon\zeta} U^{\zeta\delta} \end{pmatrix} \delta\varphi^{\epsilon} (\mathbf{r}) \\
&= 3 O_{\alpha\delta} U_{\beta}{}^{\alpha} \int d\mathbf{r} \begin{pmatrix} \left( \delta^{\delta}_{\epsilon} U^{\iota\beta} + \delta^{\beta}_{\epsilon} U^{\delta\iota} \right) \varphi_{\iota} (\mathbf{r}) \\ -\varphi^{\delta} (\mathbf{r}) \, O_{\epsilon\zeta} U^{\zeta\beta} \\ -\varphi^{\beta} (\mathbf{r}) \, O_{\epsilon\zeta} U^{\zeta\delta} \end{pmatrix} \delta\varphi^{\epsilon} (\mathbf{r}) \\
&= 3 U_{\beta}{}^{\alpha} \int d\mathbf{r} \begin{pmatrix} \left( \delta^{\gamma}_{\alpha} U^{\iota\beta} + O^{\beta\gamma} U_{\alpha}{}^{\iota} \right) \varphi_{\iota} (\mathbf{r}) \\ -\varphi_{\alpha} (\mathbf{r}) \, U^{\gamma\beta} - \varphi^{\beta} (\mathbf{r}) \, U^{\gamma}{}_{\alpha} \end{pmatrix} \delta\varphi_{\gamma} (\mathbf{r}) \\
&= 3 U_{\beta}{}^{\alpha} \begin{pmatrix} U^{\iota\beta} \langle \varphi_{\iota} | \delta\varphi_{\alpha} \rangle + U_{\alpha}{}^{\iota} \langle \varphi_{\iota} | \delta\varphi^{\beta} \rangle \\ -\langle \varphi_{\alpha} | \delta\varphi_{\gamma} \rangle U^{\gamma\beta} - \langle \varphi^{\beta} | \delta\varphi_{\gamma} \rangle U^{\gamma}{}_{\alpha} \end{pmatrix} \\
&= 3 U_{\beta}{}^{\alpha} \left( \langle \delta\varphi_{\alpha} | \varphi_{\gamma} \rangle - \langle \varphi_{\alpha} | \delta\varphi_{\gamma} \rangle \right) U^{\gamma\beta} \\
&+ 3 U_{\beta}{}^{\alpha} U_{\alpha}{}^{\gamma} \left( \langle \varphi_{\gamma} | \delta\varphi^{\beta} \rangle - \langle \delta\varphi_{\gamma} | \varphi^{\beta} \rangle \right) = 0.
\end{aligned} \tag{7.67}
$$

This confirms that the anisotropy is a tensorial invariant, remaining unchanged to first order in the Hubbard projectors. While this scalar is not, therefore, useful as a criterion to be extremised with respect to the Hubbard projectors, this result shows that the anisotropy of the DFT + $U$ interaction tensor remains unchanged to first order within a projector self-consistency scheme if the appropriate geometric corrections are applied.

Let us pause for a moment to consider the consequences of these results. If the Hubbard interaction tensor is initially computed for a given set of Hubbard projectors and interaction, then as long as these projectors span the same fixed correlated subspace and the interaction operator has no explicit dependence on the projectors, to first order, both the isotropic and anisotropic parts of the interaction tensor remain unchanged by subsequent alterations to the Hubbard projectors. Thus, the eigenvectors, and hence any physical interpretations of the interaction tensor, are unchanged during optimisation of the projectors (to first-order) so long as the appropriate geometric corrections are made, implying that re-computation of the $U$ tensor at each projector-update iteration is unnecessary.

$$
\begin{aligned}
\delta\bar{U} &= \delta U_{\alpha}{}^{\alpha\alpha}{}_{\alpha} \\
&= \left(\langle\delta\varphi_{\alpha}|\varphi_{\epsilon}\rangle - \langle\varphi_{\alpha}|\delta\varphi_{\epsilon}\rangle\right) U^{\epsilon\alpha\alpha}{}_{\alpha} + \left(\langle\delta\varphi^{\alpha}|\varphi_{\epsilon}\rangle - \langle\varphi^{\alpha}|\delta\varphi_{\epsilon}\rangle\right) U_{\alpha}{}^{\epsilon\alpha}{}_{\alpha} \\
&\quad + \left(\langle\delta\varphi^{\alpha}|\varphi_{\epsilon}\rangle - \langle\varphi^{\alpha}|\delta\varphi_{\epsilon}\rangle\right) U_{\alpha}{}^{\alpha\epsilon}{}_{\alpha} + \left(\langle\delta\varphi_{\alpha}|\varphi_{\epsilon}\rangle - \langle\varphi_{\alpha}|\delta\varphi_{\epsilon}\rangle\right) U_{\alpha}{}^{\alpha\alpha\epsilon} \\
&= 2\left(\langle\delta\varphi_{\alpha}|\varphi_{\epsilon}\rangle - \langle\varphi_{\alpha}|\delta\varphi_{\epsilon}\rangle\right) U^{\epsilon\alpha\alpha}{}_{\alpha} + 2\left(\langle\delta\varphi^{\alpha}|\varphi_{\epsilon}\rangle - \langle\varphi^{\alpha}|\delta\varphi_{\epsilon}\rangle\right) U_{\alpha}{}^{\epsilon\alpha}{}_{\alpha} \\
&= 2\left(\langle\delta\varphi_{\alpha}|\varphi_{\epsilon}\rangle - \langle\varphi_{\alpha}|\delta\varphi_{\epsilon}\rangle\right) U^{\epsilon\alpha\alpha}{}_{\alpha} + 2\left(\langle\delta\varphi^{\alpha}|\varphi_{\epsilon}\rangle - \langle\varphi^{\alpha}|\delta\varphi_{\epsilon}\rangle\right) U^{\epsilon}{}_{\alpha}{}^{\alpha}{}_{\alpha} \\
&= 2\left(\langle\delta\varphi_{\beta}|\varphi_{\epsilon}\rangle - \langle\varphi_{\beta}|\delta\varphi_{\epsilon}\rangle\right)\left(\delta^{\beta}_{\alpha} U^{\epsilon\alpha\alpha}{}_{\alpha} + O^{\beta\alpha} U^{\epsilon}{}_{\alpha}{}^{\alpha}{}_{\alpha}\right) \\
&= 2\left(\langle\delta\varphi_{\beta}|\varphi_{\epsilon}\rangle - \langle\varphi_{\beta}|\delta\varphi_{\epsilon}\rangle\right)\left(\delta^{\beta}_{\alpha}\delta^{\alpha}_{\gamma} + O^{\beta\alpha} O_{\alpha\gamma}\right) U^{\epsilon\gamma\alpha}{}_{\alpha}.
\end{aligned}
\tag{7.73}
$$

Here we have considered only real-valued functions and Hermitian operators and thereby have exploited the symmetry properties

$$
U_{\alpha}{}^{\beta\gamma}{}_{\delta} = U_{\alpha}{}^{\delta\gamma}{}_{\beta} = U_{\gamma}{}^{\beta\alpha}{}_{\delta} \quad and \quad U_{\alpha}{}^{\beta\gamma}{}_{\delta} = U^{\beta}{}_{\alpha\delta}{}^{\gamma}. \tag{7.74}
$$

While $\delta\bar{U}$ is generally non-vanishing and may serve as a measure of the magnitude of the diagonal components of the interaction tensor, allowing an equation of motion to be set up for nonorthogonal Hubbard projectors, in principle, the four-index contraction is not a valid tensorial operation and the resulting $\bar{U}$ is not invariant under basis transformations.

Another possible non-invariant measure of the quality of the Hubbard projectors, in this case one which is minimised for projectors giving an optimally spherically-symmetric density–density interaction, is the squared anisotropy of the non-invariant two-index Coulomb interaction matrix $\bar{U}_{(\alpha\beta)}$, that is

$$
\begin{aligned}
\left(\Delta\bar{U}\right)^2 &= \frac{3}{2}\sum_{\alpha\beta} U_{(\alpha\beta)} U_{(\beta\alpha)} - \frac{1}{2}\sum_{\alpha}\left(U_{(\alpha\alpha)}\right)^2 \\
&= \frac{3}{2}\sum_{\alpha\beta} U_{\alpha}{}^{\beta\alpha}{}_{\beta} U_{\beta}{}^{\alpha\beta}{}_{\alpha} - \frac{1}{2}\sum_{\alpha}\left(U_{\alpha}{}^{\alpha\alpha}{}_{\alpha}\right)^2.
\end{aligned}
\tag{7.75}
$$

In general, we would not promote the use of non-invariant scalars for measuring the Coulomb interaction when nonorthogonal Hubbard projectors are used. Put simply, the scalar given by the doubly-traced Coulomb interaction

$$
U = \sum_{\alpha,\beta} U_{\alpha}{}^{\beta\alpha}{}_{\beta} \tag{7.76}
$$

is tensorial invariant by construction, while its individual diagonal and off-diagonal contributions

$$
U' = \sum_{\alpha,\beta=\alpha} U_{\alpha}{}^{\beta\alpha}{}_{\beta} = \sum_{\alpha,\beta=\alpha} \bar{U}_{(\alpha\beta)} \quad \text{and} \tag{7.77}
$$

$$U''' = \sum_{\alpha,\beta\neq\alpha} U_{\alpha}{}^{\beta\alpha}{}_{\beta} = \sum_{\alpha,\beta\neq\alpha} \bar{U}_{(\alpha\beta)} \tag{7.78}$$

are generally not. Of course, no such impediment to their use exists in the case of orthonormal Hubbard projectors.

## 7.4 Concluding Remarks

We have discussed two approaches for determining the Hubbard $U$ parameters for ab initio methods such as DFT + $U$, in the tensorial formalism needed for nonorthogonal Hubbard projectors and in the density-matrix language of a linear-scaling DFT method. Unfortunately, neither method can be said to be straightforwardly implementable with linear-scaling cost for materials in which the number of correlated sites increases with the total number of atoms.

In the regime of very large system size, or of a large number of correlated sites, when dynamical correlation effects are important or when the implementation of another many-body perturbation theory method, such as *GW*, is of interest, we would tend to advocate the cRPA method. However, difficulties may arise for a linear-scaling implementation particularly in the numerically challenging matrix inversion required for the construction of the Green's function, which needs to be carried out once at each frequency sampled at, and the inversion of the dielectric function, which needs to be carried out separately both for each correlated site and each at frequency value. Moreover, this method has the disadvantage that an accurate set of support functions, or NGWFs, must be computed for a large number of states above the Fermi level.

On the other hand, the linear density-response method has the advantage that it depends on the accuracy of the valence bands only, and no conduction-band support functions need to be optimised. This method is ideally suited to systems with a small number of correlated sites which are significant to functionality, for example transition-metal binding sites in a biological enzyme or magnetic impurities in a conventional semiconductor, in which case linear-scaling cost may be effectively achieved. A possible technical obstacle may be encountered when the diagonal elements of the unperturbed subspace density-matrices are close to unity, where the stiffness of the response may make evaluation of the inverse response tensor difficult. Somewhat ironically, this is more likely to present a challenge for Hubbard projectors which are better suited to their chemical environment.

We propose a scheme in which both the Hubbard projectors, as in Chap. 3 and Ref. [10], and the corresponding Hubbard $U$ tensor, as in self-consistent extensions of the linear density-response [3] or cRPA [5] methods, are both self-consistently determined, in a complementary fashion whereby the projectors and parameters are re-optimised on alternate steps. Only with self-consistency between the correlated subspaces and the associated Hubbard $U$ parameters, of course, may localised correction methods such as the DFT + $U$ approximation be said to be truly ab initio and free from extraneous assumptions.

In practice, it is not immediately clear that such a non-linear system would not be under-determined, that a self-consistent solution for the density, Hubbard projectors and $U$ tensors would be the unique solution (that is unique up to arbitrary linear transformations among the Hubbard projectors for each subspace, with concomitant corrections to the $U$ tensors). Moreover, in spite of the well-behaved convergence of the Hubbard projectors with a fixed $U$, as demonstrated in Chap. 3 and Ref. [10], it may prove to be necessary to apply a mixing scheme to either the $U$ or Hubbard projectors in order to avoid oscillatory behaviour in a fully self-consistent method.

If these technical obstacles can be overcome, it is expected that the geometric approach we have described in this chapter for updating the interaction parameters automatically with the Hubbard projectors, as opposed carrying out their expensive re-calculation at each step, may provide a useful acceleration to the proposed fully self-consistent DFT + $U$ method.

## References

1. W.E. Pickett, S.C. Erwin, E.C. Ethridge, Reformulation of the LDA+U method for a local-orbital basis. Phys. Rev. B **58**(3), 1201 (1998)
2. M. Cococcioni, S. de Gironcoli, Linear response approach to the calculation of the effective interaction parameters in the LDA+U method. Phys. Rev. B **71**(3), 035105 (2005)
3. H.J. Kulik, M. Cococcioni, D.A. Scherlis, N. Marzari, Density functional theory in transition-metal chemistry: a self-consistent Hubbard $U$ approach. Phys. Rev. Lett **97**(10), 103001 (2006)
4. F. Aryasetiawan, M. Imada, A. Georges, G. Kotliar, S. Biermann, A.I. Lichtenstein, Frequency-dependent local interactions and low-energy effective models from electronic structure calculations. Phys. Rev. B **70**(19), 195104 (2004)
5. K. Karlsson, F. Aryasetiawan, O. Jepsen, Method for calculating the electronic structure of correlated materials from a truly first-principles LDA+U scheme. Phys. Rev. B **81**(24), 245113 (2010)
6. O. Gunnarsson, O.K. Andersen, O. Jepsen, J. Zaanen, Density-functional calculation of the parameters in the Anderson model: application to Mn in CdTe. Phys. Rev. B **39**(3), 1708 (1989)
7. V.I. Anisimov, O. Gunnarsson, Density-functional calculation of effective Coulomb interactions in metals. Phys. Rev. B **43**(10), 7570 (1991)
8. K. Nakamura, R. Arita, Y. Yoshimoto, S. Tsuneyuki, First-principles calculation of effective onsite Coulomb interactions of 3d transition metals: constrained local density functional approach with maximally localized Wannier functions. Phys. Rev. B **74**(23), 235113 (2006)
9. F. Aryasetiawan, K. Karlsson, O. Jepsen, U. Schönberger, Calculations of Hubbard $U$ from first-principles. Phys. Rev. B **74**(12), 125106 (2006)
10. D.D. O'Regan, N.D.M. Hine, M.C. Payne, A.A. Mostofi, Projector self-consistent DFT+U using nonorthogonal generalized Wannier functions. Phys. Rev. B **82**(8), 081102 (2010)
11. J.P. Perdew, R.G. Parr, M. Levy, J.L. Balduz, Density-functional theory for fractional particle number: derivative discontinuities of the energy. Phys. Rev. Lett **49**(23), 1691 (1982)
12. A.J. Cohen, P. Mori-Sanchez, W. Yang, Insights into current limitations of density functional theory. Science **321**(5890), 792 (2008)
13. J.F. Janak, Proof that $\frac{\partial e}{\partial n_i} = \epsilon_i$ in density-functional theory. Phys. Rev. B **18**(12), 7165 (1978)

14. E. Runge, E.K.U. Gross, Density-functional theory for time-dependent systems. Phys. Rev. Lett **52**(12), 997 (1984)
15. F. Sottile, Response functions of semiconductors and insulators: from the Bethe-Salpeter equation to time-dependent density functional theory. Ph.D. thesis, École Polytechnique, Palaiseau, France, 2003
16. M.M. Rieger, L. Steinbeck, I. White, H. Rojas, R. Godby, The GW space–time method for the self-energy of large systems. Comput. Phys. Commun. **117**(3), 211 (1999)
17. M. Springer, F. Aryasetiawan, Frequency-dependent screened interaction in Ni within the random-phase approximation. Phys. Rev. B **57**(8), 4364 (1998)
18. F. Aryasetiawan, J.M. Tomczak, T. Miyake, R. Sakuma, Downfolded self-energy of many-electron systems. Phys. Rev. Lett **102**(17), 176402 (2009)
19. R.W. Godby, P.G. González, Density Functional Theories and Self-Energy Approaches. In: C. Fiolhais, F. Nogueira, M.A.L. Marques (eds) A Primer in Density Functional Theory of Lecture Notes in Physics, vol. 620, (Springer, Heidelberg, 2003)
20. E.K.U. Gross, E. Runge, O. Heinonen, Many-Particle Theory. (Adam Hilger, Bristol, 1991)
21. L. Hedin, New method for calculating the one-particle Green's function with application to the electron-gas problem. Phys. Rev **139**, A796 (1965)
22. F. Aryasetiawan, O. Gunnarsson, The GW method. Rep. Prog. Phys **61**, 273 (1998)
23. C. Friedrich, A. Schindlmayr, Many-Body Perturbation Theory: The GW Approximation. In: J. Grotendorst, S. Blügel, D. Marx (eds) Computational Nanoscience: Do It Yourself! vol of NIC Series., (John von Neumann Institute for Computing, Jülich, 2006)
24. T. Ozaki, Efficient recursion method for inverting an overlap matrix. Phys. Rev. B **64**(19), 195110 (2001)
25. M.P. Prange, J.J. Rehr, G. Rivas, J.J. Kas, J.W. Lawson, Real space calculation of optical constants from optical to x-ray frequencies. Phys. Rev. B **80**(15), 155110 (2009)
26. C.J. Pickard, F. Mauri, Nonlocal pseudopotentials and magnetic fields. Phys. Rev. Lett **91**(19), 196401 (2003)
27. D. Varsano, L.A. Espinosa-Leal, X. Andrade, M.A.L. Marques, R. di Felice, A. Rubio, Towards a gauge invariant method for molecular chiroptical properties in TDDFT. Phys. Chem. Chem. Phys **11**, 4481 (2009)
28. P. Elliott, F. Furche, K. Burke, Excited States From Time-Dependent Density Functional Theory Reviews in Computational Chemistry, (Wiley, NJ, 2009) pp. 91–165.
29. D.D. O'Regan, M.C. Payne, A.A. Mostofi, Subspace representations in ab initio methods for strongly correlated systems. Phys. Rev. B **83**(24), 245124 (2011)
30. T. Miyake, F. Aryasetiawan, Screened Coulomb interaction in the maximally localized Wannier basis. Phys. Rev. B **77**(8), 085122 (2008)

# Chapter 8
# Discussion and Conclusion

In this Dissertation, we have described a body of research which, though hopefully described in a self-contained and internally-consistent way, so far as possible, necessarily takes root in a long and distinguished tradition of methodological development in ab initio simulation. With such a tradition in mind and a body of accompanying literature which is by now so vast, one worries about whether sufficient note has been given of any number of seminal works which have lead towards this modest one. We can only offer our apologies if this has been the case, any omission is due to our ignorance and, of course, not deliberate intent.

It has been our goal, from the outset, to grow the metaphorical tree a little bit wider, as well as taller, to make connections between different approaches and to open some fresh avenues for research. We close by offering a brief description of some ideas for future research suggested by this work. First, however, we briefly summarise what we consider to be the central conclusions of this dissertation.

## 8.1 Synopsis

In Chap. 1, we provided an introduction to ab initio simulation with Kohn–Sham DFT, with a particular emphasis on linear-scaling methods. Special focus was placed on the methodology surrounding the density-matrix formulation of Kohn–Sham DFT, namely the introduction of support functions, known NGWFs when these functions are optimised, transformations and auxiliary density kernels used to maintain density-matrix idempotency, basis sets and short-ranged numerical Fourier transforms for differential operators.

We began the discussion of strongly-correlated systems, in Chap. 2, first by introducing some pre-requisite physical concepts, the inadequacy of approximate exchange-correlation functional for such systems and the popular and efficacious DFT + $U$ method. We then detailed our linear-scaling implementation of this latter technique in the ONETEP code. In particular, we offered original derivations of the tensorially consistent DFT + $U$ contributions to the gradient terms needed to optimise



D. D. O'Regan, *Optimised Projections for the Ab Initio Simulation of Large and Strongly Correlated Systems*, Springer Theses, DOI: 10.1007/978-3-642-23238-1_8,

the electronic density and atomic positions with linear-scaling cost. In performance tests on nano-clusters of the correlated insulator nickel oxide, we demonstrated the preservation of linear-scaling performance with only a minimal increase in computational effort.

In Chap. 3, we analysed the dependence of the DFT + $U$ ground-state properties of the iron porphyrin molecule on the spatial localisation of a hydrogenic set of Hubbard projectors, quantified by an effective nuclear charge parameter $Z$. We found that observables such as electric and magnetic dipole moments and binding affinities can be very strongly affected by the choice of projectors, rendering localised self-interaction correction methods somewhat ambiguous, and their results difficult to reproduce, unless the precise procedure used to construct the Hubbard projectors is stated. A novel scheme was introduced that brings the electronic states into self-consistency with their concomitant Wannier functions. We demonstrated the behaviour of this projector self-consistent DFT + $U$ method on the iron porphyrin system, where it offers favourable results.

The salient point of Chap. 4 is that, on grounds of tensorial inconsistency, we do not concur with the most comprehensive formulation currently available in the literature for defining the projections over correlated subspaces using nonorthogonal Hubbard projectors. This seemingly innocuous question is, in fact, one which has caused a fair degree of dissent and confusion in the literature, as we discussed. By insisting on the tensorial invariance of all scalars computed with such a projection, we were led to propose an alternative formulation which, as well as being the first derived on the grounds of rigorous tensor calculus to our knowledge, has the distinct advantage of introducing potentials and forces which act locally only to subspaces on which they are explicitly dependent. This renders it naturally compatible with corrections for localised correlation effects, such as DFT + $U$, and for implementation in a linear-scaling method. We demonstrated the effect of our projection method, by comparing it to the foremost alternative, in a projector self-consistent DFT + $U$ study of both bulk nickel oxide and the copper phthalocyanine dimer.

Chapter 5 detailed original research into the geometric ramifications of the use of a set of nonorthogonal support functions to represent the single-particle density matrix. This geometric theory allowed us to address a long-standing issue in linear-scaling ab initio methods, namely the density-matrix purity preservation under support function optimisation. This development potentially allows for many further advances to be made, not alone in the energy minimisation problem, but also, for example, in the optimisation of projections for strongly-correlated subspaces and in the time-propagation of the density-matrix for large systems.

We applied this geometric methodology to the energy minimisation algorithm of ONETEP in Chap. 6, describing a detailed numerical study on selected hydrocarbon polymers. In this study, we verified the principal prediction of the geometric theory, that is the density-matrix invariance under support function optimisation. We also noted a number of other advantages, in terms of numerical stability, and suggested ways in which those gains may be exploited.

In Chap. 7, we discussed the challenge of computing the Hubbard interaction tensor $U$ with linear-scaling expense. Two ab initio methods for determining $U$ were described, generalised to the case of nonorthogonal projectors and cast into the tensorial formalism of Chap. 4, namely the linear density-response technique and the constrained Random Phase Approximation. We then drew upon the geometric arguments of Chap. 5 in order to investigate how this tensor should change with the Hubbard projectors, suggesting a way in which to accelerate the performance of a combination of the projector self-consistency technique of Chap. 3 with a method for determining the Hubbard $U$.

## 8.2 Future Work

Finally, we briefly mention some promising directions for future research which are motivated by this dissertation.

The geometric corrections of Chap. 5 have been implemented in a straightforward manner in ONETEP and no effort has been made to adapt the energy minimisation algorithm to take special advantage of the purity of the density-matrix after support function update. At present, purity checks and safety mechanisms are in place which are thought to be counterproductive and unnecessary when the geometric corrections are applied.

It remains to be seen under what circumstances these mechanisms may be obviated and if any boost to performance may be achieved. We are only beginning to see the benefit of this technique, and, in particular, it will be interesting to apply it to some systems for which energy minimisation has previously proved difficult. Moreover, we have yet to implement the strategy for adapting the conjugate-gradients algorithm for the curvature of the support manifold as proposed in Appendix A.3; this stands as an intriguing direction for future investigation.

It will, of course, be necessary to apply our linear-scaling implementation of DFT + $U$ to some large systems for the purposes of extracting scientific predictions and further testing our methodological developments. Technologically interesting candidates for study include defective bulk transition-metal or lanthanoid oxides, particularly where those defects carry a magnetic moment, surfaces of such substances, or open-shell defect ions in otherwise simple semiconductor crystals. We are currently engaging in separate collaborative activities for each of these three system-types.

In the study of such systems, we would be eager to make use of the projector self-consistent DFT + $U$ technique, of Chap. 3 where possible, since this method seems to bring us some way towards removing the ambiguity with respect to the choice of correlated subspaces. Furthermore, its cost is not prohibitive. We would favour the use of the localised "tensorial" representation of Chap. 4 in such investigations. While we have implemented the DFT + $U$ forces using conventional hydrogenic projectors, it would be worthwhile to also do this for the self-consistent Hubbard projectors in the form of NGWFs, either by re-optimising the projector set at each ionic configuration step or simply spatially translating them after an initial optimisation. Of course, this is likely to be rather more challenging in terms of computing resources.

One of our main priorities, moving forward, is to provide a robust implementation of a first-principles method for computing the Hubbard $U$ tensor. The linear density-response method, summarised in Chap. 7, is promising in this regard since it appears to be relatively simple to implement, conceptually transparent and free of artefacts due to the proximity of periodic images in the size regime appropriate to a linear-scaling method.

The geometric correction terms discussed in Chap. 7 may perhaps permit us to reduce the cost of re-computing such interaction tensors when the Hubbard projectors are altered, by allowing us to compute a first-order estimate of its change instead. However, it remains to be seen if this approach is beneficial in practice. It is our ultimate goal, notwithstanding, to concatenate the calculation of both the Hubbard projectors, and the corresponding $U$ tensor, in a fully self-consistent DFT + $U$ method for unambiguous calculation of the properties of large strongly-correlated systems.

# Appendix
# Geometric Observations

In this appendix, we offer comments on some geometric aspects of the the optimisation of nonorthogonal support functions, additional to those given in Chap. 5. In particular, we give a formal geometrical justification of the gradient correction technique of Refs. [1, 2], together with an estimate of the error incurred; a characterisation of the differential geometry of the support manifold; and a simple adaptation of the Fletcher–Reeves non-linear conjugate gradients method for application to curved manifolds.

## A.1 Geometric Error Estimate for Energy Gradients

The metric compatibility of the support manifold has important consequences for facilitating methods in which the support functions are updated, for example in order to minimise the total energy. In this particular case, one most easily computes the direction opposite to the derivative of the energy with respect to the support functions (noting that no geometric corrections are needed to transport a scalar invariant), so that the contravariant search direction is given by

$$g^{\alpha} = -\frac{\partial E}{\partial \phi_{\alpha}}, \tag{A.1}$$

which is itself a contravariant vector since

$$\begin{aligned} \phi_{\alpha}(\tilde{\mathbf{r}}) &= \frac{\partial}{\partial \tilde{\mathbf{r}}} \phi_{\alpha}(\mathbf{r}) \quad \Rightarrow \\ g^{\alpha}(\tilde{\mathbf{r}}) &= -\frac{\partial E}{\partial \phi_{\alpha}(\tilde{\mathbf{r}})} = -\frac{\partial \tilde{\mathbf{r}}}{\partial \mathbf{r}} \frac{\partial E}{\partial \phi_{\alpha}(\mathbf{r})} = \frac{\partial \tilde{\mathbf{r}}}{\partial \mathbf{r}} g^{\alpha}(\mathbf{r}). \end{aligned} \tag{A.2}$$

In order to preserve the tensorial character of the support functions however, and to compute update directions which are indeed going to take the energy in its

D. D. O'Regan, *Optimised Projections for the Ab Initio Simulation of Large and Strongly Correlated Systems*, Springer Theses, DOI: 10.1007/978-3-642-23238-1, 

direction of steepest descent, we may make use of the metric compatibility of the support manifold in order to provide a search direction which transforms as a covariant vector. The necessity of providing gradient steps of the correct tensor character has been noted and addressed previously in the context of electronic structure calculations in two seminal works on the topic, namely Refs. [1, 2].

The central point is that we can only add quantities with identical tensor character while retaining this character. For example, as we have seen in Chap. 4, published in Ref. [3], we cannot symmetrise a tensor over pairs of indices with mixed character while generally preserving a tensor. Typically, returning to our example of an energy gradient, the metric tensor computed at the point at which the gradient is calculated is used to provide, approximately, a covariant search direction, so the support functions are updated according to

$$\phi^1_\alpha = \phi_\alpha + \lambda g_\alpha = \phi_\alpha + \lambda g^\beta S_{\beta\alpha}, \tag{A.3}$$

where $\lambda$ is an appropriate step length.

Considering partial derivatives of a real-valued metric tensor only, we might conclude that the error in the metric induced by this approximation at the point $\phi^1_\alpha$ is given by

$$S^1_{\alpha\beta} - S_{\alpha\beta} = 2\lambda\langle g_\alpha|\phi_\beta\rangle, \tag{A.4}$$

so that the error in the covariant step introduced by neglecting changes to the metric tensor during the line step is thus on the order of

$$\lambda g^\beta(S^1_{\beta\alpha} - S_{\beta\alpha}) = 2\lambda^2|g^\beta\rangle\langle g_\beta|\phi_\alpha\rangle \approx \mathcal{O}(\lambda^2). \tag{A.5}$$

This is not a complete picture, however, since it ignores the constraint that the updated support functions must lie in the cotangent space of the support manifold at some point. In order to properly estimate the error in the covariant step, we must include the geometric corrections due to the Christoffel symbols in the first covariant derivative of the metric. As we have previously shown, the absolute derivative of the metric tensor vanishes (Ricci's Lemma), so that error in the metric tensor vanishes to first order in the step length. As a result, we find that the error is smaller than we might naively expect, since in fact

$$\lambda g^\beta(S^1_{\beta\alpha} - S_{\beta\alpha}) \approx \lambda g^\beta \mathcal{O}(\lambda^2) \approx \mathcal{O}(\lambda^3). \tag{A.6}$$

The neglect of the change in the metric tensor upon support function update is thus quite fully justified.

The commutativity of the metric with covariant differentiation also allows us to compute a dimensionless angular quantifier on changes to the set of support functions. The angle $\theta$ between two unit vectors, $\Psi$ and $\Phi$, defined at a point on the support manifold, is given by the familiar formula

$$\cos(\theta) = \Psi_\alpha\Phi^\alpha = \Phi_\alpha S^{\alpha\beta}\Phi_\beta, \tag{A.7}$$

where the metric tensor is evaluated at that point. This generalises to non-unit vectors by the renormalised expression

$$\cos(\theta) = \frac{\Psi_\alpha \Phi^\alpha}{\sqrt{|\Psi_\beta \Psi^\beta||\Phi_\gamma \Phi^\gamma|}} = \frac{\Psi_\alpha S^{\alpha\beta} \Phi_\beta}{\sqrt{|\Psi_\beta S^{\beta\delta} \Psi_\delta||\Phi_\gamma S^{\gamma\epsilon} \Phi_\epsilon|}}. \tag{A.8}$$

Consequently, the angle between two vectors $\phi$ and a nearby vector $\phi + \Delta\phi$ (also a unit vector since it is complete, by definition), which span the cotangent spaces at two nearby points on the support manifold, is given by

$$\begin{aligned} \Delta\theta &= \arccos\left(\frac{\phi_\alpha(\phi^\alpha + \Delta\phi^\alpha)}{\sqrt{|\phi_\beta \phi^\beta|}\sqrt{|(\phi_\gamma + \Delta\phi_\gamma)(\phi^\gamma + \Delta\phi^\gamma)|}}\right) \\ &= \arccos\left(\frac{N - |\phi_\alpha \Delta\phi^\alpha|}{\sqrt{N}\sqrt{N}}\right) \\ &= \arccos\left(1 - \frac{|\phi_\alpha \Delta\phi^\alpha|}{N}\right) \\ &\approx \sqrt{2\varepsilon} + \mathcal{O}\left(\varepsilon^{3/2}\right) \quad \left(\text{where } \varepsilon = \frac{|\phi_\alpha \Delta\phi^\alpha|}{N} \ll 1\right), \end{aligned} \tag{A.9}$$

where $N$ is the total number of support functions.

This provides an alternative quantifier of changes in the support functions to a simply computation of the geodesic length of change vector, i.e.,

$$\Delta l = \sqrt{\Delta\phi_\alpha \Delta\phi^\alpha}. \tag{A.10}$$

The ratio of the two measures, that is

$$r = \frac{\Delta l}{\Delta\theta} = \sqrt{\frac{N\Delta\phi_\alpha \Delta\phi^\alpha}{2\phi_\beta \Delta\phi^\beta}}, \tag{A.11}$$

gives an estimate of the radius of geodesic curvature of the support manifold in the support function search direction $\Delta\phi_\bullet$.

## A.2 Differential Curvature of the Support Manifold

Let us next explore some of the geometric properties of the support manifold generating complete sets of support functions, the Levi-Civita connection on which is described in Sect. 5.4. A central element of Riemannian geometry is the skew-symmetric Riemann-Christoffel tensor defined as

$$R^\alpha_{\beta\gamma\delta} = \Gamma^\alpha_{\delta\beta,\gamma} - \Gamma^\alpha_{\gamma\beta,\delta} + \Gamma^\alpha_{\gamma\epsilon}\Gamma^\epsilon_{\delta\beta} - \Gamma^\alpha_{\delta\epsilon}\Gamma^\epsilon_{\gamma\beta}, \tag{A.12}$$

which is a commutator on second covariant differentiation. For an example of its utility, for mixed second derivatives of a rank-1 covariant tensor $A_\bullet$ we have that

$$R^{\alpha}_{\beta\gamma\delta}A_\alpha = A_{\beta,\gamma\delta} - A_{\beta,\delta\gamma}. \tag{A.13}$$

For the geometry under investigation here, we obtain that

$$\begin{aligned}\Gamma^{\alpha}_{\delta\beta,\gamma} &= -\frac{\partial}{\partial\phi^\gamma}\left[\phi^\alpha S_{\delta\beta}\right] = -\delta^\alpha_\gamma S_{\delta\beta} - \phi^\alpha S_{\delta\beta,\gamma} \quad \text{and}\\ \Gamma^{\alpha}_{\gamma\epsilon}\Gamma^{\epsilon}_{\delta\beta} &= \phi^\alpha S_{\gamma\epsilon}\phi^\epsilon S_{\delta\beta} = \delta^\alpha_\gamma S_{\delta\beta}.\end{aligned} \tag{A.14}$$

The appropriate Reimann–Christoffel tensor is thus provided by

$$\begin{aligned}R^{\alpha}_{\beta\gamma\delta} &= -\delta^\alpha_\gamma S_{\delta\beta} - \phi^\alpha S_{\delta\beta,\gamma} + \delta^\alpha_\delta S_{\gamma\beta} + \phi^\alpha S_{\gamma\beta,\delta} + \delta^\alpha_\gamma S_{\delta\beta} - \delta^\alpha_\delta S_{\gamma\beta}\\ &= -\phi^\alpha S_{\delta\beta,\gamma} + \phi^\alpha S_{\gamma\beta,\delta}\\ &= \langle\phi^\alpha|\phi_\beta\rangle S_{\delta\gamma} + \langle\phi^\alpha|\phi_\delta\rangle S_{\beta\gamma} - \langle\phi^\alpha|\phi_\beta\rangle S_{\gamma\delta} - \langle\phi^\alpha|\phi_\gamma\rangle S_{\beta\delta}\\ &= \delta^\alpha_\delta S_{\beta\gamma} - \delta^\alpha_\gamma S_{\beta\delta},\end{aligned} \tag{A.15}$$

which implies that the support manifold is not a flat space in general. Furthermore, this result shows that second covariant differentiation is not commutative, and in fact that

$$A_{\beta,\gamma\delta} - A_{\beta,\delta\gamma} = S_{\beta\gamma}A_\delta - S_{\beta\delta}A_\gamma. \tag{A.16}$$

Although we do not explore the issue of non-commutativity in mixed second absolute derivatives further, similar expressions may be derived for higher-order tensors. In particular, it may be necessary to consider mixed second derivatives of the density kernel, for example, when computing the optimal step length in a non-linear conjugate gradients scheme optimised for the non-trivial curvature of support manifold. Such a method lies beyond the scope of this study and remains as a future avenue for investigation.

Notwithstanding, we may make some further observations on the geometry of the support manifold. In particular, the fully covariant (known as the curvature tensor) and fully contravariant versions of the Riemann-Christoffel tensor are given, respectively, by

$$\begin{aligned}R_{\alpha\beta\gamma\delta} &= S_{\alpha\delta}S_{\beta\gamma} - S_{\alpha\gamma}S_{\beta\delta} \quad \text{and}\\ R^{\alpha\beta\gamma\delta} &= S^{\alpha\delta}S^{\beta\gamma} - S^{\alpha\gamma}S^{\beta\delta}.\end{aligned} \tag{A.17}$$

A related and important quantity, known as the Ricci curvature tensor, is in this case

$$R_{\alpha\beta} = R^{\gamma}_{\alpha\gamma\beta} = \delta^\gamma_\beta S_{\alpha\gamma} - \delta^\gamma_\gamma S_{\alpha\beta} = -(N-1)S_{\alpha\beta}. \tag{A.18}$$

An Einstein manifold such as this one is a special case of a Riemannian manifold (one whose Ricci curvature is a scalar multiple of its metric tensor) [4].

A number of curvature invariants may be built from the Riemann-Christoffel tensor, the simplest of which is the linear invariant given by the contraction

$$R = R_{\alpha\beta}S^{\beta\alpha} = -(N-1)S_{\alpha\beta}S^{\beta\alpha} = -N(N-1). \tag{A.19}$$

The simplest way in which to measure the curvature of the support manifold is to employ the *sectional curvature*, which is defined at each point on the surface in terms of two test vectors in the tangent space, denote them $\Psi^{\bullet}$ and $\Phi^{\bullet}$. The curvature at this point with respect to the plane formed by these two vectors (or any linear combination of the pair) is given by the scalar invariant

$$K = \frac{R_{\alpha\beta\gamma\delta}\Psi^{\alpha}\Psi^{\gamma}\Phi^{\beta}\Phi^{\delta}}{\left(S_{\alpha\gamma}S_{\delta\beta} - S_{\alpha\delta}S_{\gamma\beta}\right)\Psi^{\alpha}\Psi^{\gamma}\Phi^{\beta}\Phi^{\delta}}. \tag{A.20}$$

In our case, we see that the curvature becomes independent of both the test vectors and, as a result (known as Schur's theorem [5]), independent also of the point at which it is evaluated. Such a space of constant curvature

$$K = -1 \tag{A.21}$$

is known as a *hyperbolic manifold* of unit radius (this may also be established from Eq. A.19), or a Gauss-Bolyai-Lobachevsky space. Hyperbolic spaces enjoy a rich set of properties which are, in a sense, opposite to those of spherical surfaces. The internal angles of triangles made up of geodesic lines sum to less than $\pi$, where they would sum to greater than $\pi$ for hyperspherical surfaces. Lines which are at one point parallel tend to drift apart in hyperbolic surface, instead of towards each other as in the case of spherical surfaces. Generally, the hyperbolic plane may be thought of as one which contorts in such a manner so that no self-intersection occurs, i.e., so that no closed geodesic lines are allowed, in contrast to spherical surfaces which curve inwards to form infinitely many closed geodesic lines.

The geometric properties of the support manifold are surely interesting and may, with further investigation, yield support function algorithms which are optimised to take advantage of such properties as manifold curvature, as we describe in the following section. We refer the reader to Ref. [6], in particular, for an important work on geometry-adapted energy minimisation algorithms. We have limited ourselves in this work, however, to making the best of the conventional conjugate gradients algorithm (albeit with non-linear trial steps) available in the ONETEP method, investigating the effect of geometric corrections to the density kernel within that framework in Chap. 6.

## A.3 Geometrically-Modified Conjugate Gradients Algorithm

Finally, we briefly elucidate the point concerning the contribution to gradient conjugacy deviation from support manifold curvature made in Sect. 6.3, proposing a simple corrective step to approximately remove it. We have not performed this correction to the Fletcher–Reeves non-linear conjugate gradients algorithm in the calculations described in Chap. 6, however, for to do so would further complicate the analysis of the effect of the geometric corrections to the density kernel. A numerical investigation remains as a future avenue for investigation.

Let us suppose that the sum of the history of gradients (suppressing the step lengths for simplicity) including that at step $i-1$, and valid in the NGWF representation at that step, is given by

$$\tilde{G}_{\alpha}^{(i-1)} = \sum_{j=1}^{i-1} g_{\alpha}^{(j)}. \tag{A.22}$$

The difference between the NGWFs at step $i-1$ and the current step $i$ is denoted $\Delta\phi_{\alpha}$, we have computed the gradient $g_{\alpha}^{(i)}$ and we wish to take its inner product with $\tilde{G}_{\alpha}^{(i-1)}$. Strictly speaking, we may not straightforwardly do so, unfortunately, because $\tilde{G}_{\alpha}^{(i-1)}$ is the valid representation only in the frame of NGWFs at step $i-1$. In order to bring the gradient history $\tilde{G}_{\alpha}^{(i-1)}$ up to date, we must consider the absolute derivative of the gradients with respect to the NGWFs, that is given by

$$\tilde{G}_{\alpha}^{;\beta} = \tilde{G}_{\alpha,\gamma} S^{\gamma\beta} - \Gamma^{\delta}_{\alpha\gamma} \tilde{G}_{\delta} S^{\gamma\beta}. \tag{A.23}$$

Of course, since the history of gradients has no explicit dependence on what we subsequently do to the NGWFs, there is no partial derivative term when we evaluate this expression at $\tilde{G}_{\alpha}^{(i-1)}$, however the term proportional to a Christoffel symbol is generally non-zero. The gradient history appropriate to the current frame of NGWFs, call it $G_{\alpha}^{(i-1)}$, thus may be computed from that in the previous frame using the expression

$$\begin{aligned} G_{\alpha}^{(i-1)} &= \tilde{G}_{\alpha}^{(i-1)} + \langle \Delta\phi_{\beta} | \tilde{G}_{\alpha}^{(i-1);\beta} \rangle \\ &= \tilde{G}_{\alpha}^{(i-1)} - \langle \Delta\phi_{\beta} | \Gamma^{\delta}_{\alpha\gamma} \tilde{G}_{\delta}^{(i-1)} S^{\gamma\beta} \rangle \\ &= \tilde{G}_{\alpha}^{(i-1)} + \langle \Delta\phi_{\beta} | \phi^{\delta} S_{\alpha\gamma} \tilde{G}_{\delta}^{(i-1)} S^{\gamma\beta} \rangle \\ &= (\delta_{\alpha}{}^{\beta} + \langle \Delta\phi_{\alpha} | \phi^{\beta} \rangle) \tilde{G}_{\beta}^{(i-1)}. \end{aligned} \tag{A.24}$$

As a result, the conjugacy condition for a geometrically adapted conjugate gradients algorithm is not the conventional expression, $\langle g^{(i)\alpha} | \tilde{G}_{\alpha}^{(i-1)} \rangle$, but rather the modified form

$$\langle g^{(i)\alpha} | G_{\beta}^{(i-1)} \rangle = \langle g^{(i)\alpha} | \tilde{G}_{\beta}^{(i-1)} \rangle (\delta^{\beta}{}_{\alpha} + \langle \phi^{\beta} | \Delta\phi_{\alpha} \rangle). \tag{A.25}$$

This final expression provides a computationally inexpensive first-order correction to the conjugate gradients algorithm for support function nonorthogonality.

# References


1. E. Artacho, L. Miláns del Bosch, Nonorthogonal basis sets in quantum mechanics: representations and second quantization. Phys. Rev. A **43**(11), 5770 (1991)
2. C.A. White, P. Maslen, M.S. Lee, M. Head-Gordon, The tensor properties of energy gradients within a non-orthogonal basis. Chem. Phys. Lett. **276**(1−2), 133 (1997)
3. D.D. O'Regan, M.C. Payne, A.A. Mostofi, Subspace representations in ab initio methods for strongly correlated systems. Phys. Rev. B **83**(24), 245124 (2011)
4. B. Spain, *Tensor Calculus—A Concise Course* (Dover, Spain, 2003)
5. D. Lovelock, H. Rund, *Tensors, Differential Forms and Variational Principles* (Dover, Spain, 1989)
6. A. Edelman, T.A. Arias, S.T. Smith, The geometry of algorithms with orthogonality constraints. SIAM J. Matrix Anal. Appl. **20**(2), 303 (1998)

# Curriculum Vitae

## David D. O'Regan

Post-doctoral Research Associate, Theory of Condensed Matter Group, Cavendish Laboratory 19, J. J. Thomson Ave., University of Cambridge, CB3 0HE, UK.
E-mail: ddo20@cam.ac.uk Tel: +447938873417, Birth/Nationality: Cork, Ireland.

## Education

| | |
|---|---|
| 2007–2011 | Ph.D., Theoretical Physics (successfully examined 15/4/11), University of Cambridge. *On optimised projections for the ab initio simulation of large and strongly-correlated systems* (supervised by Professor Mike C. Payne F.R.S.). |
| 2003–2007 | University College Cork (National University of Ireland): BSc. Joint Honours Mathematics (First Class) and Physics (First Class). Final year mean: Mathematics 75%, Physics 86%. |
| 1997–2003 | Irish Leaving Certificate, 555 points: English (A1); Chemistry (A1); Mathematics (A2); Physics (A2); French (A2); Irish (B1); Applied Maths. (C2), Technical Drawing (OA1). |

## Recent Publications and Presentations

D. D. O'Regan, M. C. Payne and Arash A. Mostofi, *Subspace representations in ab initio methods for strongly correlated systems.* Phys. Rev. B 83, 245124 (2011).

D. D. O'Regan, *Optimised Projections for the Ab Initio Simulation of Large and Strongly Correlated Systems*, Springer Theses, DOI: 10.1007/978-3-642-23238-1,

D. D. O'Regan, N. D. M. Hine, A. A. Mostofi and M. C. Payne, *Projector self-consistent DFT+U using nonorthogonal generalized Wannier functions*. Phys. Rev. B 82, 081102(R) (2010)—An "Editor's suggestion".

D. D. O'Regan, *Geometric corrections for nonorthogonal representation optimisation in linear-scaling DFT*. Article to be submitted (2011).

D. D. O'Regan, N. D. M. Hine, M. C. Payne and A. A. Mostofi, *Linear-scaling DFT+U energies and forces for large, strongly correlated systems*. Letter in preparation.

D. D. O'Regan, M. Robinson, N. D. M. Hine, A. A. Mostofi and P. D. Haynes, *Linear-scaling calculations of dielectric properties using nonorthogonal generalised Wannier functions*. Article in preparation.

D. D. O'Regan, *Optimised projections and tensorial invariance in methods for large and strongly-correlated systems*. Oral presentation, Materials Modelling Laboratory, Oxford, UK, June 2011.

D. D. O'Regan, N. D. M. Hine, M. C. Payne and A. A. Mostofi, *Projector self-consistent DFT+U with linear-scaling and tensorial invariance*. Poster Presentation, ACAM workshop, Dublin, Ireland, June 2011.

S. M.-M. Dubois, L. Ratcliff, D.D. O'Regan et al., *Prospects for linear scaling calculations beyond DFT accuracy*. Poster Presentation, CECAM Workshop on GW Calculations, Lausanne, Switzerland, June 2011.

D. D. O'Regan, *Decorrelating user and theory in ab initio calculations on strongly correlated materials: Projector self-consistent DFT+U for large systems*. Oral presentation and invited participant, CECAM Workshop on Materials Informatics, Lausanne, Switzerland, May 2011.

D. D. O'Regan, *Nonorthogonal Wannier function optimisation: geometric aspects and linear-scaling applications to strongly-correlated materials and spectroscopy*, Invited DIPC seminar, University of the Basque Country, San Sebastian, Spain, May 2011.

D. D. O'Regan, A. A. Mostofi and M. C. Payne, *Linear-scaling and projector self-consistent DFT+U for electronic correlations in large systems. Poster presentation, Psi_k Conference 2010*, Berlin, Germany, September 2010; Thomas Young Centre International Review, University College London, UK, September 2010.

D. D. O'Regan, M. C. Payne and A. A. Mostofi, *Linear-scaling and projector self-consistent DFT+U for electronic correlations in large systems*. Oral presentation, March 2010 meeting of American Physical Society, Portland, OR, USA.

D. D. O'Regan. *Accelerating linear-scaling DFT: Differential geometry meets electronic structure theory*. Oral presentations, Cavendish Laboratory, November 2010; Imperial College London, UK, December 2010.

D. D. O'Regan, A. A. Mostofi and M. C. Payne. *Studying molecular magnets with ONETEP*. Oral presentation, Thomas Young Centre, Imperial College London, UK, October 2009.

D. D. O'Regan, A. A. Mostofi and M. Payne. *Wannier functions in the study of strong electronic correlations in DNA. Poster presentation, International*

*Workshop on DNA-based nanotechnology: Construction, mechanics, and electronics*, Max-Planck-Institut für Physik komplexer Systeme, Dresden, Germany, May 2009.

D. D. O'Regan, A. A. Mostofi and M. C. Payne, *Linear scaling DFT+U with ONETEP, Poster presentation, 4th International Workshop on Computational Physics and Materials Science: Total Energy and Force Methods*, Trieste, Italy. CCP9 conference, Cambridge, UK. Sept. 2008.

## Teaching

| | |
|---|---|
| Summer 2011 | Summer project supervisor (optical spectra of defects in diamond), Cavendish Laboratory. |
| Spring 2010 | Undergraduate final year project supervisor (topic of DFT+U), Cavendish Laboratory. |
| 2008–2010 | Supervisor in Mathematics for Natural Scientists, Pembroke College, Cambridge. |
| 2007–2009 | Demonstrator for Theoretical Physics II (3rd year course), Cavendish Laboratory. |

## Awards, Professional Activities and Meetings

EPSRC Post-doctoral Research Fellowship, February–September 2011.

National University of Ireland Travelling Studentship in the Sciences, 2007.

EPSRC Ph.D. studentship, 2007–2010. Cambridge University European Trust Honorary Scholarship, 2007.

Awarded "College Scholar" of the University College Cork in each year of undergraduate study.

Scholarship to attend "Spins in Solids" Physics Summer School, June 2006, University of Virginia.

Reviewer for Europhysics Letters.

Academic visitor to Thomas Young Centre (Mostofi Group), Imperial College London, 2007–present.

Prominent contributor to ONETEP (including DFT+U, TDDFT, geometrical density-matrix preservation).

Co-ordinator of regular "Density Functional Theory users and developers meeting", Cavendish Laboratory.

Associate Member of Institute of Physics, former IoP Nexus representative for Ireland.

Former UCC Physics and Astronomy Society recording secretary and treasurer.

Scholarship awarded, attended 4th *International Workshop and School on Time-Dependent Density Functional Theory*, Centro de Ciencias de Benasque Pedro Pascual, Spain, January 2010.
Scholarship awarded, CECAM Summer School *Theoretical Spectroscopy Lectures*, Zurich, Switzerland, May 2009.
Attended *Physics by the Lake* EPSRC summer school, Ambleside, UK, August, 2008.
Talks on ONETEP software contributions, *ONETEP summer school*, Cambridge, U.K., July 2008 and April 2010.
Attended Computational Molecular Science, Bristol, UK, June 2008.

## Previous Research Activities

| | |
|---|---|
| January–April 2007 | Final Research project at NUI Cork Mathematics Dept. *Differential Geometry of the Clockmaker's Fusèe*, awarded 1st Class Honours. |
| June–September 2006 | Science Foundation Ireland grant to work at the Tyndall National Institute, Cork City, under Prof. Stephen B. Fahy. Computational research project in FORTRAN, *Carrier interaction with concentration fluctuations in semiconductor alloys.* |
| June–August 2005 | Employee of Surface Science Division, College of Optical Sciences, University of Arizona, USA under Professor Charles M. Falco. Characterisation of epitaxially-grown thin magnetic thin-films for novel magnetic data-processing technology. |

## Additional Skills

| | |
|---|---|
| Languages: | Irish, French, Spanish. OS: Unix, Linux, Mac OSX, etc. |
| Programming: | Expertise in high performance computing and parallelisation, FORTRAN, Python, LaTeX. |
| Data analysis: | Mathematica, gnuplot, Xmgrace, XcrySDen, jmol, vmd, Python, shell scripting. |
| Ab initio software: | ONETEP, CASTEP, PWscf, Opium, Wannier90, Octopus, some Abinit. |

## Leisure Activities

Restoration and preservation of vintage motorcycles, member Munster Vintage Motorcycle and Car Club. Live music, linguistics, political science, cookery, cycling, swimming, travel, ski, photography.

## Research Interests

Optimised Wannier function and density-matrix methods, associated tensor calculus and numerical analysis.
Methods for strong correlation (DFT+U and MBPT) and optical (particularly chiroptical) spectra (TDDFT).
Wannier function formulation of Modern Theories, response functions and methods for finite applied fields.
Close collaboration with experiment and new avenues for physically-motivated development of theory.

## CV Effective June 2011, with Referees

Professor Mike C. Payne F.R.S.
Head of Research Group, Theory of Condensed Matter Group,
Cavendish Laboratory, 19, J. J. Thomson Avenue,
Cambridge CB3 0HE, UK.
mcp1@cam.ac.uk

Dr. Arash A. Mostofi
Senior Lecturer & RCUK Fellow, Department of Materials,
Imperial College London, Exhibition Road,
London SW7 2AZ, UK.
a.mostofi@imperial.ac.uk

Professor Stephen B. Fahy, SFI Principal Investigator,
Condensed Matter Theory, Tyndall National Institute,
Lee Maltings, Cork City, Ireland.
stephen.fahy@tyndall.ie

MIX
Papier aus verantwortungsvollen Quellen
Paper from responsible sources
FSC® C105338

FSC
www.fsc.org

If you have any concerns about our products,
you can contact us on
**ProductSafety@springernature.com**

In case Publisher is established outside the EU,
the EU authorized representative is:
**Springer Nature Customer Service Center GmbH**
**Europaplatz 3, 69115 Heidelberg, Germany**

Printed by Libri Plureos GmbH
in Hamburg, Germany